6 SBAC Math Practice Tests Grade 7

Extra Practice to Achieve a Crack Score

By

Elise Baniam & Michael Smith

6 SBAC Math Practice Tests Grade 7

Published in the United State of America By

The Math Notion

Email: info@mathnotion.com

Web: www.mathnotion.com

ISBN: 978-1-63620-160-3

About the Author

Elise Baniam has been a math instructor for over a decade now. She graduated in Mathematics. Since 2006, Elise has devoted his time to both teaching and developing exceptional math learning materials. As a Math instructor and test prep expert, Elise has worked with thousands of students. She has used the feedback of her students to develop a unique study program that can be used by students to drastically improve their math score fast and effectively.

- **ACT Math Workbook**
- **SAT Math Workbook**
- **ISEE/SSAT Math Workbook**
- **Common Core Math Workbook**
- **many Math Education Workbooks**
- **and some Mathematics books ...**

As an experienced Math teacher, Mrs. Baniam employs a variety of formats to help students achieve their goals: she teaches students in large groups, and she provides training materials and textbooks through her website and through Amazon.

You can contact Elise via email at:
Elise@mathnotion.com

6 Practice Tests to Help Achieve an Excellent SBAC Math Score!

Practice makes perfect, and the best way to exercise your SBAC test-taking skills is with simulated tests. Our experts selected these targeted questions to help you study more realistically and use your review time wisely to reach your best score. These math questions are the same as the ones you will find on the SBAC test, so you will know what to expect and avoid surprises on test day.

6 SBAC Math Practice Tests Grade 7 provides six full-length opportunities to evaluate whether you have the skills to ace the test's higher-level math questions.

This book emphasizes that any difficult math question focuses on building a solid understanding of basic mathematical concepts. Inside the practice math book, you will find realistic SBAC math questions and detailed explanations to help you master your math sections of the SBAC. You will discover everything you need to ace the test, including:

- Fully explained answers to all questions.
- **Aligned to State and National Standards.**
- Practice questions that help you increase speed and accuracy.
- Learn fundamental approaches for achieving content mastery.
- Diagnose and learn from your mistakes with in-depth answer explanations.

With the SBAC math prep, the lots of students who would like an intensive drill with multiple math questions, get a quick but full review of everything on their exam. Anyone planning to take the SBAC exam should take advantage of math practice tests. Purchase it today to receive access to 7th grade SBAC math practice questions.

WWW.MATHNOTION.COM

… So Much More Online!

- ✓ FREE Math Lessons
- ✓ More Math Learning Books!
- ✓ Mathematics Worksheets
- ✓ Online Math Tutors

For a PDF Version of This Book

Please Visit www.mathnotion.com

Contents

SBAC Mathematics Test Review

Bubble Answer Sheet

Remove (or photocopy) the answer sheet and use it to complete the practice tests.

Name: **Date:**

1	Ⓐ Ⓑ Ⓒ Ⓓ	16	Ⓐ Ⓑ Ⓒ Ⓓ
2	Ⓐ Ⓑ Ⓒ Ⓓ	17	Ⓐ Ⓑ Ⓒ Ⓓ
3	Ⓐ Ⓑ Ⓒ Ⓓ	18	Ⓐ Ⓑ Ⓒ Ⓓ
4	Ⓐ Ⓑ Ⓒ Ⓓ	19	Ⓐ Ⓑ Ⓒ Ⓓ
5	Ⓐ Ⓑ Ⓒ Ⓓ	20	Ⓐ Ⓑ Ⓒ Ⓓ
6	Ⓐ Ⓑ Ⓒ Ⓓ	21	Ⓐ Ⓑ Ⓒ Ⓓ
7	Ⓐ Ⓑ Ⓒ Ⓓ	22	Ⓐ Ⓑ Ⓒ Ⓓ
8	Ⓐ Ⓑ Ⓒ Ⓓ	23	Ⓐ Ⓑ Ⓒ Ⓓ
9	Ⓐ Ⓑ Ⓒ Ⓓ	24	Ⓐ Ⓑ Ⓒ Ⓓ
10	Ⓐ Ⓑ Ⓒ Ⓓ	25	Ⓐ Ⓑ Ⓒ Ⓓ
11	Ⓐ Ⓑ Ⓒ Ⓓ	26	Ⓐ Ⓑ Ⓒ Ⓓ
12	Ⓐ Ⓑ Ⓒ Ⓓ	27	Ⓐ Ⓑ Ⓒ Ⓓ
13	Ⓐ Ⓑ Ⓒ Ⓓ	28	Ⓐ Ⓑ Ⓒ Ⓓ
14	Ⓐ Ⓑ Ⓒ Ⓓ	29	Ⓐ Ⓑ Ⓒ Ⓓ
15	Ⓐ Ⓑ Ⓒ Ⓓ	30	Ⓐ Ⓑ Ⓒ Ⓓ

SBAC Grade 7 Mathematics Reference Materials

Linear Equations			
Slope-intercept form			$y = mx + b$
Constant of proportionality			$k = \frac{y}{x}$
Circumference			
Circle	$C = 2\pi r$	or	$C = \pi d$
Area			
Triangle			$A = \frac{1}{2}bh$
Rectangle or Parallelogram			$A = bh$
Trapezoid			$A = \frac{1}{2}h(b_1 + b_2)$
Circle			$A = \pi r^2$
Volume			
Prism or cylinder			$V = Bh$
Pyramid or Cone			$V = \frac{1}{3}Bh$
Additional Information			
Pi	$\pi = 3.14$	or	$\pi = \frac{22}{7}$
Distance			$d = rt$
Simple interest			$I = prt$
Compound interest			$I = p(1 + r)^t$

Smarter Balanced Assessment Consortium

SBAC Practice Test 1

Mathematics

GRADE 7

Administered *Month Year*

1) Peter paid for 9 sandwiches.

- Each sandwich cost 12.34.
- He paid for 8 bags of fries that each cost $2.25.

Which equation can be used to determine the total amount, y, Peter paid?

A. $y = 9(12.34) + 8(2.25)x$

B. $y = (12.34 + 2.25)x$

C. $y = 9(12.34) + 8(2.25)$

D. $y = 12.34x + 8(2.25)$

2) What is the decimal equivalent of the fraction $\frac{7}{22}$?

A. 0.18

B. $0.\overline{18}$

C. $0.3\overline{18}$

D. 0.318

3) The circumference of a circle is $16\,\pi$ centimeters. What is the area of the circle in terms of π?

A. $16\,\pi$

B. $64\,\pi$

C. $32\,\pi$

D. $48\,\pi$

4) What is the volume of rectangular prism when the two triangular prisms below are stuck together?

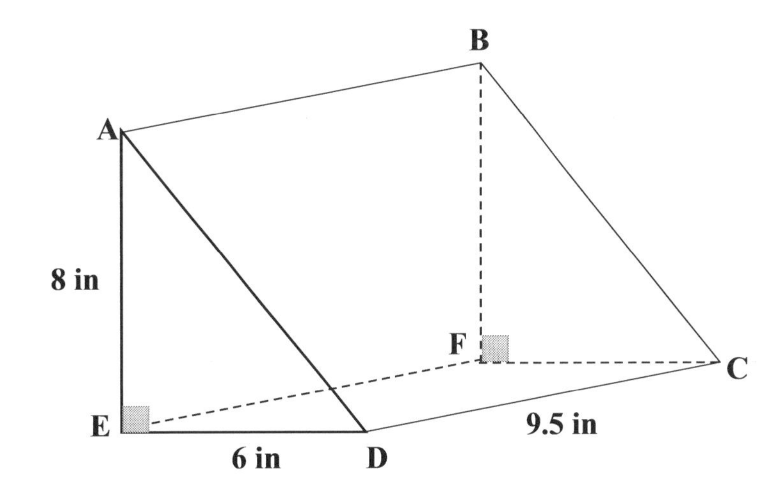

A. 456 in^3

B. 228 in^3

C. 24 in^3

D. 9.5 in^3

5) Which number line Shows the solution to the inequality $-5x - 4 < -9$?

A.

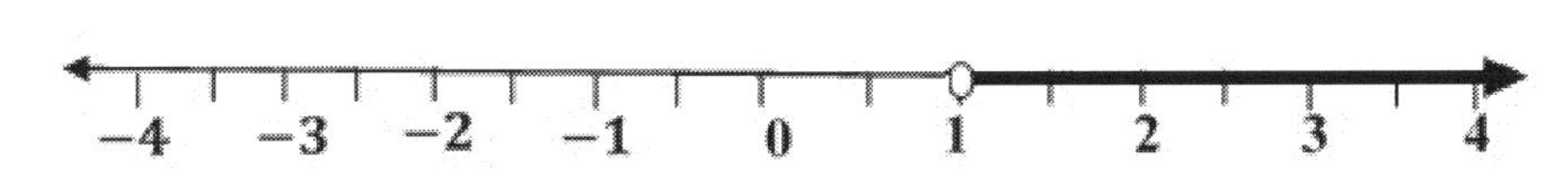

B.

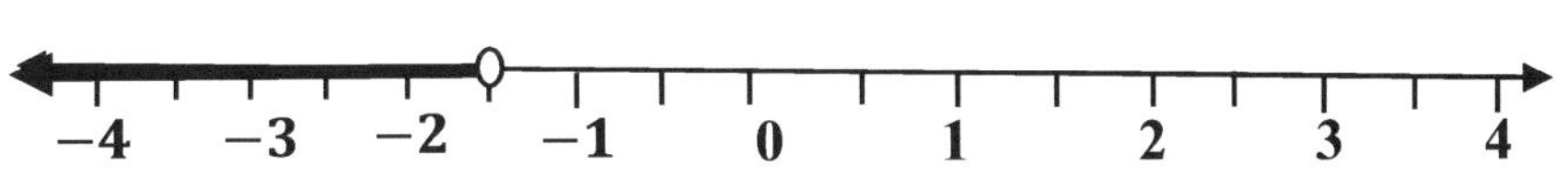

C.

D.

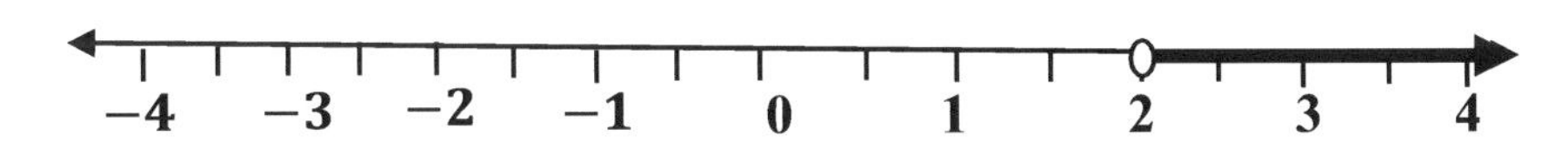

6) What is the value of $(2 + 7)^2 + (2 - 7)^2$?

A. -116

B. 86

C. 116

D. 106

7) Arsan has $16 to spend on school supplies. The following table shows the price of each item in the school store. No sale tax is charged on these items.

Which the combination of items can Arsan buy with his $16?

Item	Price
Notebook	$3.75
Pen	$0.90
Eraser	$ 1.30
Folder	$1.95

A. 6 Notebooks and 4 Pens

B. 4 Folders and 6 Erasers

C. 5 Notebooks and 2 Folders

D. 7 Erasers and 9 Pens.

8) If 17% of x is 51, what is 45% of x?

A. 135

B. 48.5

C. 30.04

D. 1.40

9) If all variables are positive, find the square root of $\frac{25x^9y^5}{49xy}$?

A. $\frac{5}{7}x^3y$

B. $\frac{5y^2}{7x^3}$

C. $\frac{5}{7}x^4y^2$

D. $5\frac{1}{7}x^4y^2$

10) Which is closcst to the perimeter of the right triangle in the figure below?

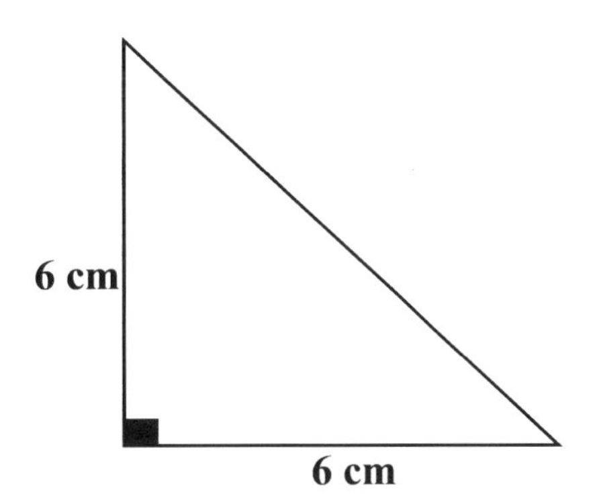

A. 22.2

B. 13

C. 15

D. 20.5

11) What is the range of the following set of data: $1, -3, 5, -2, 2, 7, 0, 2, 12$?

A. 2

B. 8

C. 12

D. 15

12) Alex starts a saving account with $80. Every week he adds $9 to his account.

Which equation can be used to determine the number of weeks w, after which

Alex's accounts reaches $192?

A. $9w + 192 = 80$

B. $9 + w = 192$

C. $9w + 80 = 192$

D. $9w - 80 = 192$

13) The medals won by United States, Australia and Spain during a basketball competition are shown in the table below:

United States	Australia	Spain
5	8	7

Out of the medals won by these three countries, what percentage of medals did the United States win?

A. 20%

B. 25%

C. 35%

D. 15%

14) A girl in State A spent $44 before a 2.5% sales tax and a girl in State B spent $40 before an 1.75% sales tax. How much more money did the girl from State A spend than the girl from State B after sales tax was applied? Round to the nearest hundredth.

A. 4.40

B. 44.70

C. 2

D. 18.7

15) A school has 432 students and 19 chemistry teachers and 16 physics teachers. What is the ratio between the number of physics teachers and the number of students at the school?

A. $\frac{1}{27}$

B. $\frac{16}{27}$

C. $\frac{1}{16}$

D. $\frac{16}{19}$

16) James has his own lawn mowing service. The maximum James charges to mow a lawn is $51. Which inequality represents the amount James could charge, P, to mow a lawn?

A. $P < 51$

B. $P = 51$

C. $P \leq 51$

D. $P \geq 51$

17) What is the value of this expression $22 \div 0.55$?

A. 0.4

B. 4.40

C. 30

D. 40

18) The ratio of boys to girls in Maria Club is the same as the ratio of boys to girls in Hudson Club. There are 24 boys and 42 girls in Maria Club. There are 20 boys in Hudson Club. How many girls are in Hudson Club?

A. 35

B. 18

C. 28

D. 35

19) On average, Simone drinks $\frac{3}{4}$ of a 5-ounce glass of coffee in $\frac{1}{4}$ hour. How much coffee does she drink in an hour?

A. 2.5 ounces

B. 1.5 ounces

C. 15 ounces

D. 25 ounces

20) Line P, R, and S intersect each other, as shown in below diagram. Based on the angle measures, what is the value of θ?

A. 29°

B. 107°

C. 78°

D. 73°

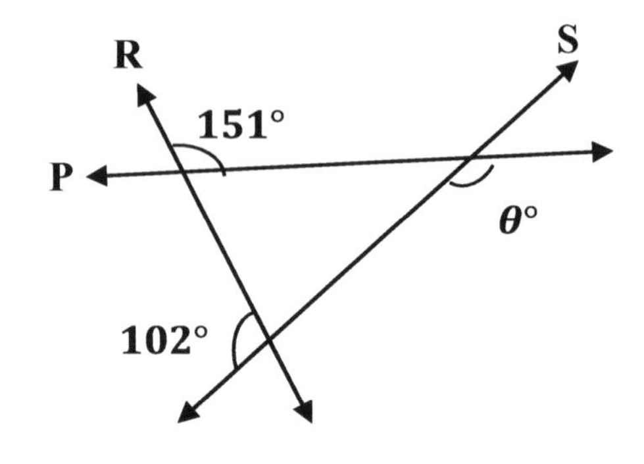

21) Which expression is represented by the model below?

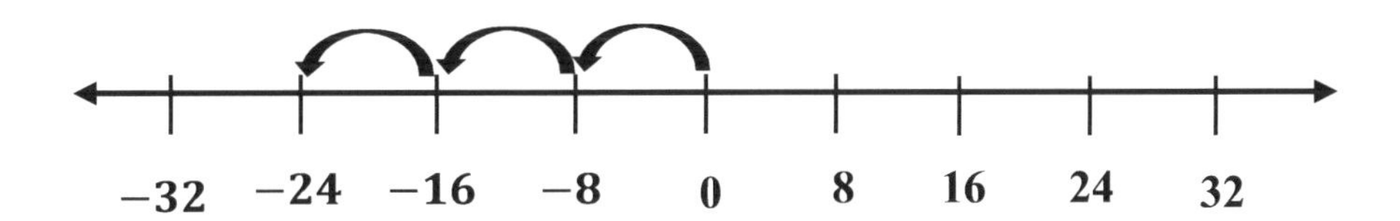

A. $-3 . (-8)$

B. $(-3) . 8$

C. $3 . (-8)$

D. $3 . 8$

22) The table below shows the distance, y, a lion can travel in mile in x hour.

Time (x, hour)	Distance (y, mile)
6	228
12	456
18	684
24	912
30	1,140

Based on the information in the table, which equation can be used to model the relationship between x and y?

A. $y = x + 6$

B. $y = 6x$

C. $y = x + 228$

D. $y = 38x$

23) Mia has a loan of $38,450. This loan has a simple interest rate of 1.8% per year. What is the amount of interest that Mia will be charged on this loan at the end of one year?

A. $169.2

B. $669.2

C. $992.6

D. $692.1

24) The spinner shown has eight congruent sections.

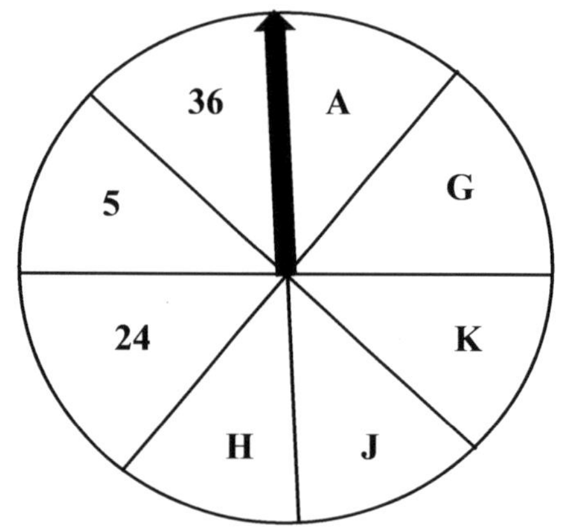

The spinner is spun 240 times. What is a reasonable prediction for the number of times the spinner will land on a letter?

A. 45

B. 90

C. 120

D. 150

25) Which graph best represents the distance a car travels when going 30 miles per hour?

A.

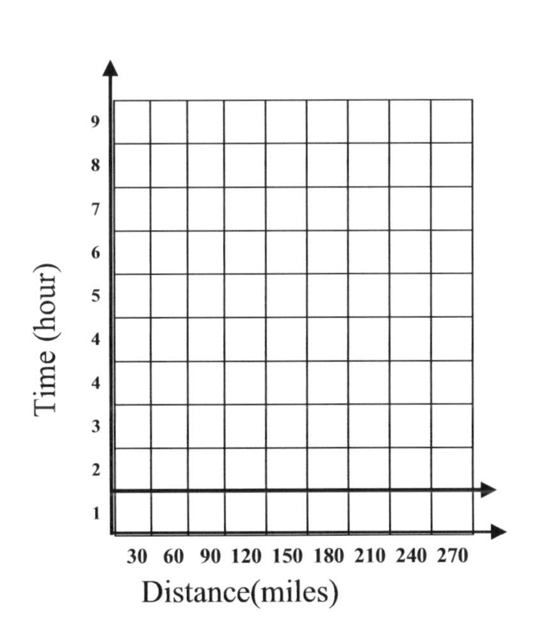

B.

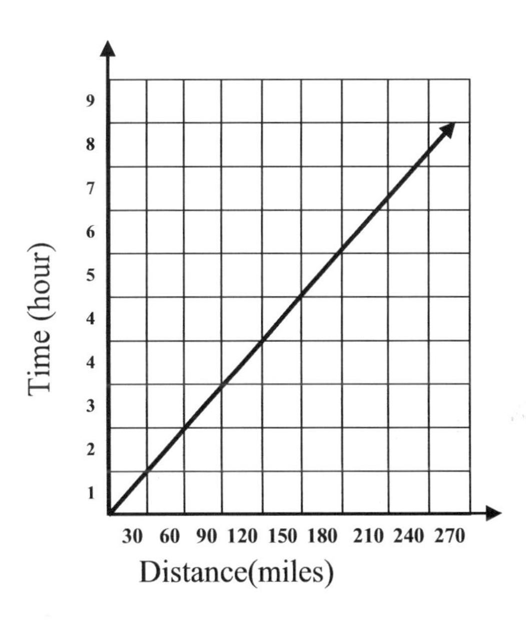

C.

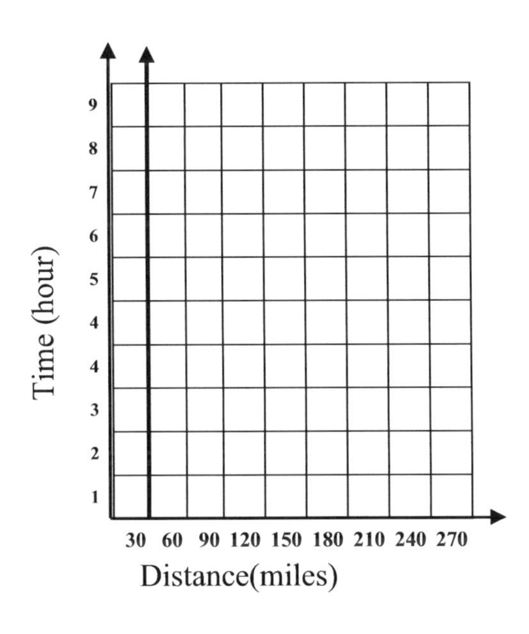

D.

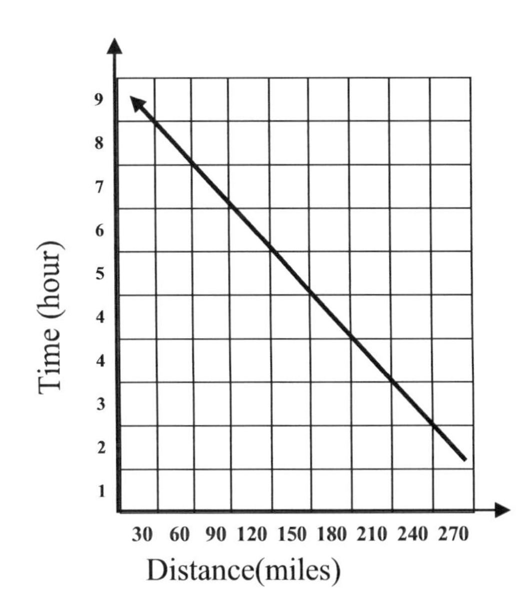

26) The temperature is shown in the table below, on each of day in the week for a city in February. What is the mean temperature, in the city for that week?

A. -14

B. -49

C. -7

D. -7.7

Day	Temperature (°F)
Monday	-24
Tuesday	-31
Wednesday	-15
Thursday	-11
Friday	0
Saturday	12
Sunday	20

27) Which arithmetic sequence is represented by the expression $6m - 5$, where m represents the position of a term in the sequence?

A. 7, 13, 19, 26, 32, …

B. 7, 13, 19, 25, 31, …

C. 13, 19, 24, 28, 31, …

D. 13, 19, 22, 25, 28, …

28) Which expression is equivalent to $-26 - 312d$?

A. $-13(2 - 34d)$

B. $-6(30d - 8)$

C. $-13(2 + 24d)$

D. $-348\ d$

29) The dot plots show how many minutes per day do 7th grade study math after school at two different schools on one day.

Number of minuets study in school 1

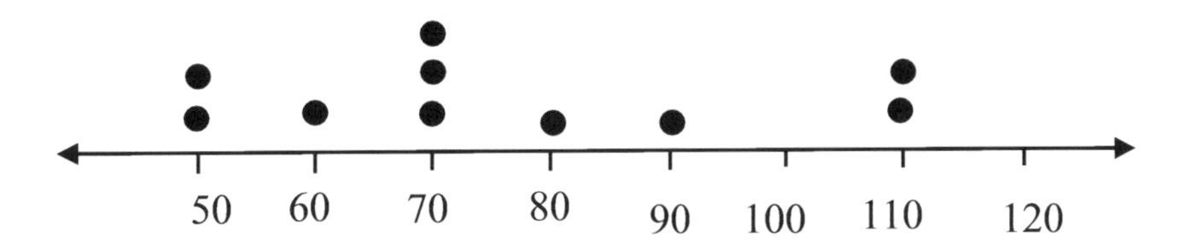

Number of minuets study in school 2

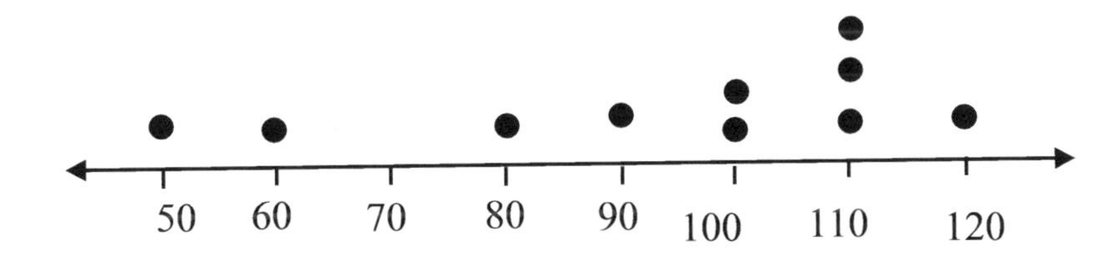

Which statement is supported by the information in the dot plots?

A. The mode of the data for School 2 is greater than the mode of the data for School 1.

B. The mean of the data for School 1 is greater than the mean of the data for School 2.

C. The median of the data for School 2 is smaller than the median of the data for School 1.

D. The median and mean of the data for two schools are equal.

30) Which number represents the probability of an event that is unlikely to occur?

A. 0.97

B. 1.6

C. 0.20

D. 0.82

Smarter Balanced Assessment Consortium

SBAC Practice Test 2

Mathematics

GRADE 7

Administered *Month Year*

1) What is the decimal equivalent of the fraction $\frac{34}{11}$?

A. $\overline{3.09}$

B. $3.0\overline{90}$

C. $3.\overline{09}$

D. 3.09

2) Thomas is shareholder of a company. The price of stock is $75.85 on the morning of day 1. Thomas records the change in the price of the stock in the chart below at the end of each day, but some information is missing.

Day	Change in Price ($)
1	+ 0.52
2	+0.66
3	
4	−0.70
5	

The change in the price for day 3 is $\frac{2}{5}$ of the change in the price for day 4. At the end of day 5, the price of Thomas's stock is $76.92. What is the change, in dollars, in the price of the stock for day 5?

A. −0.65

B. 0.87

C. −0.87

D. 0.65

3) Kevin adds $\frac{2}{9}$ cups of sugar into a mixture every $\frac{1}{5}$ hour. What is the rate, in cups per minute, at which Kevin adds sugar to the mixture?

A. $\frac{1}{54}$

B. $1\frac{1}{8}$

C. $\frac{1}{15}$

D. $\frac{1}{108}$

4) A box of ball contains 11 blue balls, 9 red balls, 10 black balls, and 3 green balls. All the balls are the same size and shape. Brian will select a ball at random. Which of the following best describes the probability that Brian will select a green ball?

A. unlikely

B. certain

C. likely

D. impossible

5) The first number in a pattern is 4. Each following number is found by subtracting 5 from the previous number. What is the sixth number in the pattern?

A. -26

B. -31

C. -16

D. -21

6) Evelyn opened a bank account. She adds the same amount of money to her account each month. The table below shows the amounts of money in her account at the ends of certain numbers of months.

How much money does Evelyn add to her bank account each month?

Month	Amount
4	$64
7	$112
10	$160

A. $15

B. $10

C. $16

D. $17

7) Using data from house sales, probabilities for the story of a house sold were calculated. The probabilities for two story are listed below.

- The probability a house sold has one story is 0.35.
- The probability a house sold has two story 0.50.

Based on these probabilities, how many of the next 400 houses sold are likely to be one story and how many are likely to be two story?

A. one story: 35, two story: 50

B. one story: 200, two story: 100

C. one story: 140, two story: 200

D. one story: 200, two story: 141

8) Mr. Turner is digging a trench to put in the new school sprinkler system. Every $\frac{1}{9}$ hour, the length of his trench increases by $\frac{2}{7}$ foot. By how much does the length, in feet, of Mr. Turner's trench increase each hour?

A. $\frac{1}{7}$

B. $\frac{3}{17}$

C. $\frac{7}{18}$

D. $\frac{18}{7}$

9) Multiply: $4\frac{3}{8} \times \frac{-3}{8}$

A. $-1\frac{7}{8}$

B. $-1\frac{1}{9}$

C. $-1\frac{41}{64}$

D. -41

10) Brendan charges $31 per hour plus $60 to enter data. He accepted a project for no more than $820. Which inequality can be used to determine all the possible numbers of hours (x) it took the man to enter the data?

A. $31x + 60 \leq 820$

B. $31x + 60 > 820$

C. $60x + 31 < 820$

D. $60x + 31 \geq 820$

11) Use the coordinate grid below to answer the question. What is the circumference of the circle?

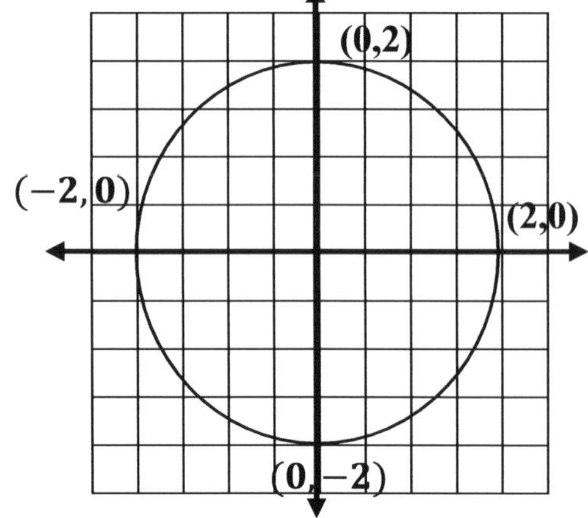

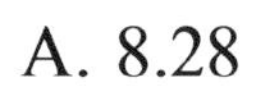

12) The temperature is 8° F. As a cold front move in, the temperature drops 4° F per half hour. What is the temperature at the end of 3 hours?

A. 24°F

B. 12°F

C. −32°F

D. −16°F

13) A printer originally cost h dollars, including tax. Eddy purchased the printer when it was on sale for 32% off its original cost. Which of the following expressions represents the final cost, in dollars, of the printer Eddy purchased?

A. $h + 0.68$

B. $h - 0.32$

C. $0.68h$

D. $0.32h$

14) Use the set of data below. What is the median of the list of numbers?

42, 21, 28, 36, 26, 21

A. 36

B. 27

C. 21

D. 42

15) Asher worked out at a gym for 3 hours. His workout consisted of jogging for 48 minutes, playing volleyball for 42 minutes, and playing billiards for the remaining amount of time. What percentage of Asher's workout was spent playing billiards?

A. 50%

B. 25%

C. 45%

D. 35%

16) The angle measures of a triangle GBD are shown in the diagram. What is the value of $\angle B$?

A. 20

B. 104

C. 85

D. 102

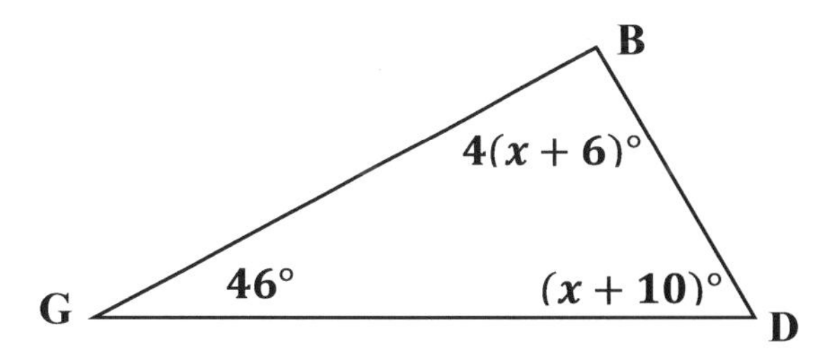

17) Triangle PRS is shown on the grid below,

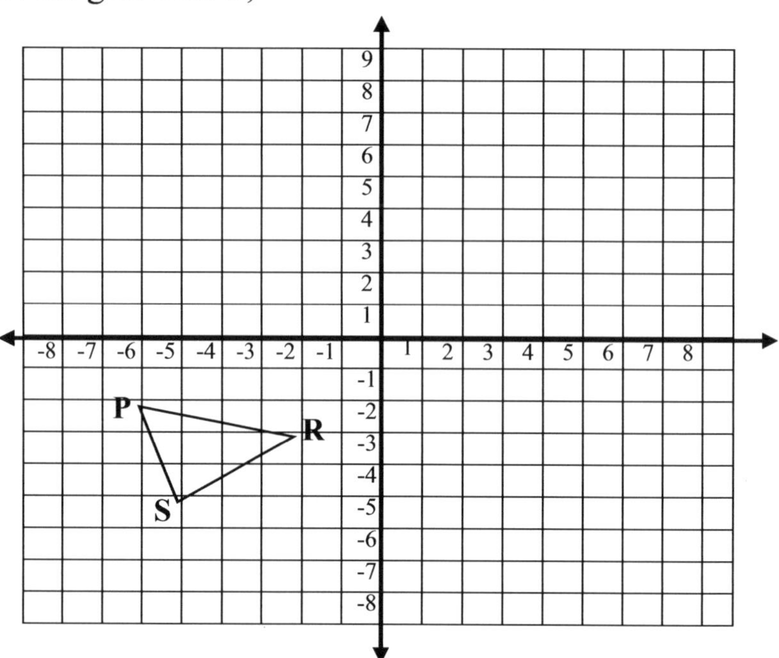

If triangle PRS is reflected across the y-axis to form triangle P′R′S′, which ordered pair represents the coordinates of P′?

A. $(-6, -2)$

B. $(6, 2)$

C. $(6, -2)$

D. $(2, -6)$

18) What is the solution set for the inequality $-5x + 12 > -8$?

A. $x > 4$

B. $x < 4$

C. $x > -4$

D. $x < -4$

19) In a party people drink 88.71 liters of juices. There are approximately 29.57 milliliters in 1 fluid ounce. Which measurement is closest to the number of fluid ounces in 118.28 liters?

A. 0.003 fl oz

B. 3,782.48 fl oz

C. 3,008.84 fl oz

D. 3,000 fl oz

20) The dimensions of a square pyramid are shown in the diagram. What is the volume of the square pyramid in cubic inches?

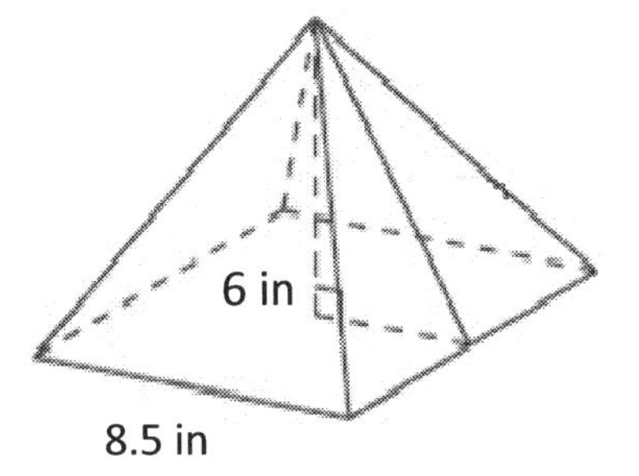

A. 144.5 in^3

B. 114.6 in^3

C. 288.2 in^3

D. 454.5 in^3

21) Water is poured to fill a pool in the shape of a rectangular prism. The pool is 12 feet long, 7.5 feet wide, and 11.8 feet high. How much cubic feet of water are needed to fill the pool?

A. 1,062 ft^3

B. 106.2 ft^3

C. 162.20 ft^3

D. 106.90 ft^3

22) The store manager spent $11,8400 to buy a new freezer and 25 tables. The total purchase is represented by this equation, where v stands for the value of each table purchased: $25v + 2{,}340 = 11{,}840$

What was the cost of each table that the manager purchased?

A. $280

B. $288

C. $308

D. $380

23) In a city, at 3:25 A.M., the temperature was $-4°F$. At 3:25 P.M., the temperature was $17°F$. Which expression represents the increase in temperature?

A. $-4 + 17$

B. $|-4 - 17|$

C. $|-4| - 17$

D. $-4 + |-17|$

24) Angles α and β are complementary angles. Angles α and are supplementary angles. The degree measure of angle β is $118°$. What is the measure of angle γ?

A. $28°$

B. $118°$

C. $62°$

D. $40°$

25) The bar graph shows a company's income and expenses over the last 5 years.

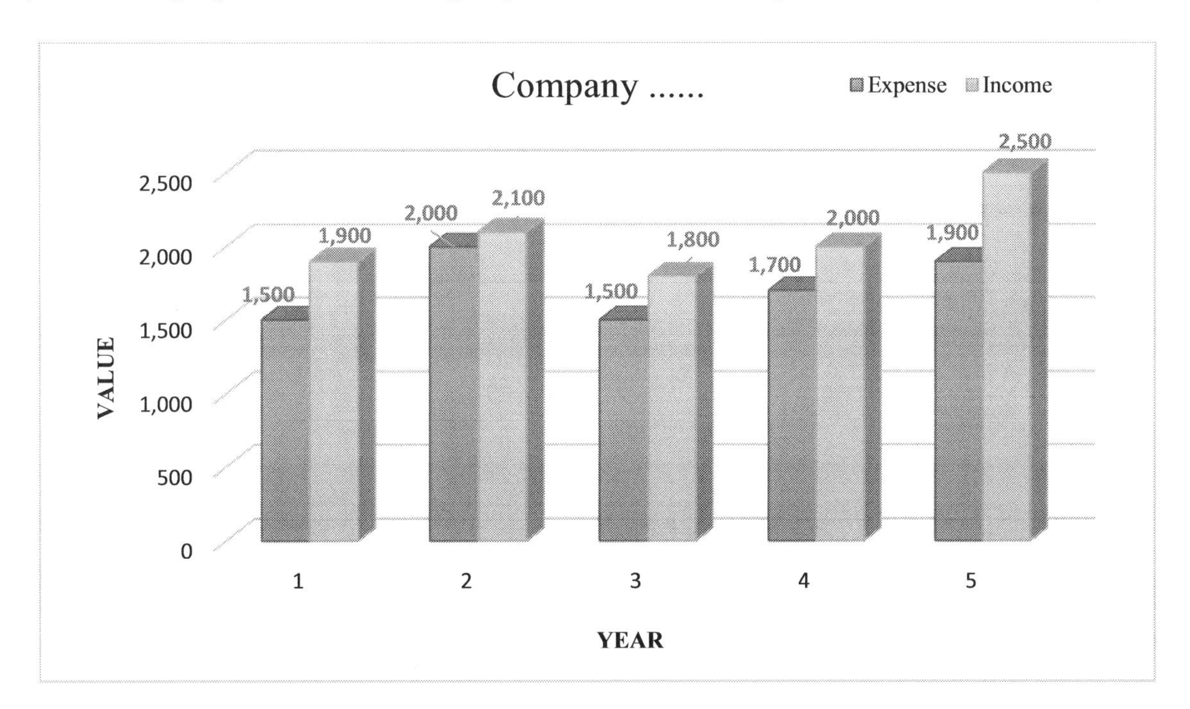

Which statement is supported by the information in the graph?

A. Expenses have increased \$500 each year over the last 5 years.

B. The income in Year 5 was 20% more than the income in Year 1.

C. The combined income in Years 3, 4, and 5 was equal to the combined expenses in Years 2, 3, and 4.

D. Expenses in the year 3 was more than half of the income in the year 4.

26) Which expression is equivalent to the $(4n - 7) - \frac{1}{2}(11 - 8n) + \frac{7}{2}$?

A. -4

B. $-4n - 9$

C. $8n - 9$

D. $8n - 4$

27) Patricia bought a bottle of 16-ounce balsamic vinegar for $12.14. She used 45% of the balsamic vinegar in two weeks. Which of the following is closest to the cost of the balsamic she used?

A. $0.54

B. $5.65

C. $5.46

D. $4.56

28) A scale drawing of triangle DEF that will be used on a wall is shown below. What is the perimeter, in meter, of the actual triangle used on the wall?

Scale: 1 cm: $2\frac{1}{3}$ m

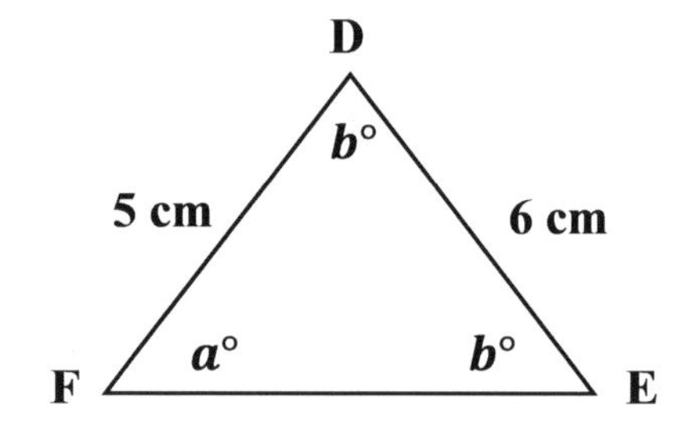

A. $41\frac{2}{3}$

B. $37\frac{3}{7}$

C. $37\frac{1}{3}$

D. $43\frac{1}{3}$

29) The ratio of boys to girls in Geometry class is 2 to 7. There are 42 girls in the class. What is the total number of students in Geometry class?

A. 52

B. 54

C. 28

D. 14

30) A group of employees have their weight recorded to make a data set. The mean, median, mode, and range of the data set are recorded. Then, the weight of the manager is included to make a new data set. The manager's weight is more than all but one of the employees. Which measure must be the same when the manager's weight included?

A. Mean

B. Mode

C. Median

D. Range

Smarter Balanced Assessment Consortium

SBAC Practice Test 3

Mathematics

GRADE 7

Administered *Month Year*

1) Peter paid for 7 sandwiches.

- Each sandwich cost 10.22.
- He paid for 6 bags of fries that each cost $1.87.

Which equation can be used to determine the total amount, y, Peter paid?

A. $y = 7(10.22) + 6(1.87)x$

B. $y = (10.22 + 1.87)x$

C. $y = 7(10.22) + 6(1.87)$

D. $y = 10.22x + 6(1.87)$

2) What is the decimal equivalent of the fraction $\frac{6}{11}$?

A. 0.54

B. $0.4\overline{54}$

C. $0.\overline{54}$

D. 0.545

3) The circumference of a circle is 18 π centimeters. What is the area of the circle in terms of π?

A. 18 π

B. 81 π

C. 36 π

D. 54 π

4) What is the volume of rectangular prism when the two triangular prisms below are stuck together?

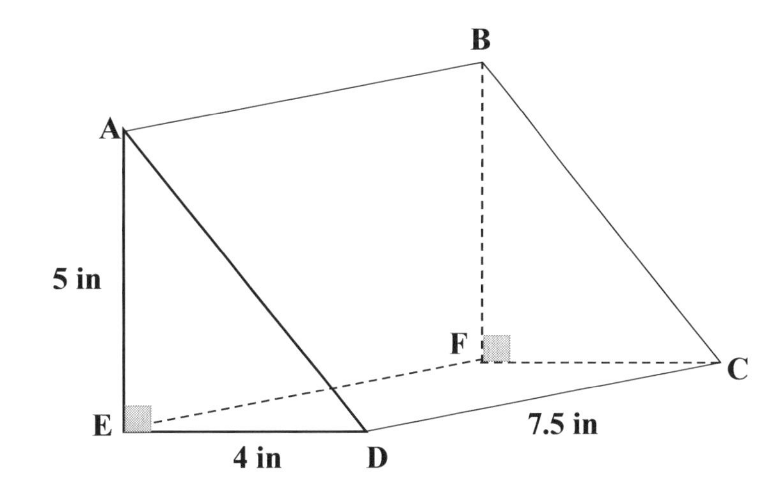

C. 56 in^3

D. 7.5 in^3

5) Which number line Shows the solution to the inequality $-7x - 3 < -10$?

A.

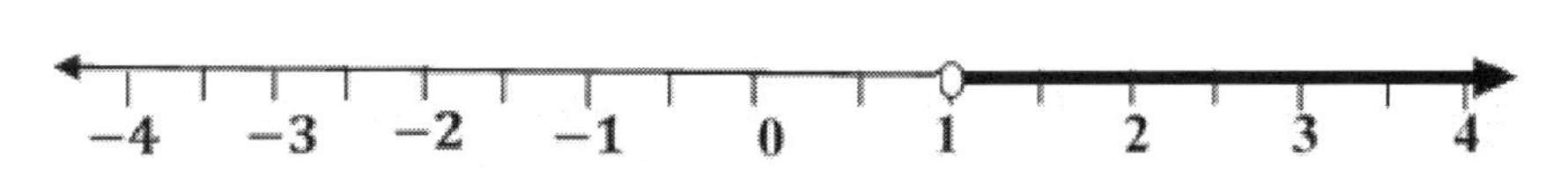

B.

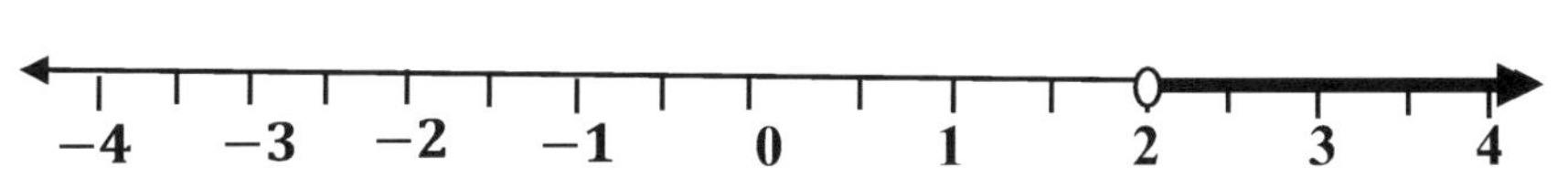

C.

D.

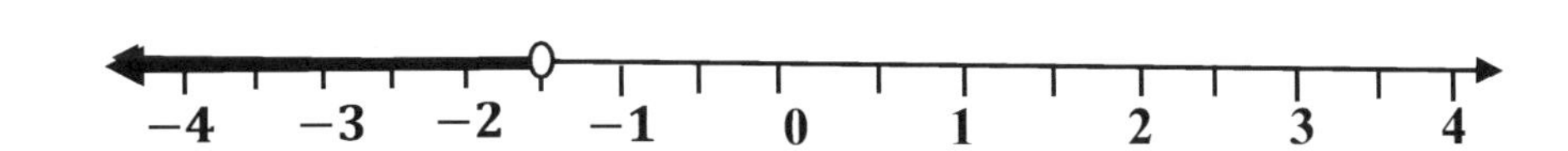

6) What is the value of $(3 + 8)^2 + (3 - 8)^2$?

A. -157

B. 25

C. 121

D. 146

7) Arsan has $11 to spend on school supplies. The following table shows the price of each item in the school store. No sale tax is charged on these items.

Which the combination of items can Arsan buy with his $11?

A. 4 Notebooks and 2 Pens

B. 3 Folders and 5 Erasers

C. 2 Notebooks and 4 Folders

D. 6 Erasers and 6 Pens.

Item	Price
Notebook	$2.25
Pen	$1.10
Eraser	$ 0.85
Folder	$2.15

8) If 18% of x is 72, what is 35% of x?

A. 140

B. 44.8

C. 14.04

D. 1.40

9) If all variables are positive, find the square root of $\frac{16x^9y^3}{81xy}$?

A. $\frac{4}{9}x^6y$

B. $\frac{9y}{4x^4}$

C. $\frac{4}{9}x^4y$

D. $4\frac{4}{5}x^3y^4$

10) Which is closest to the perimeter of the right triangle in the figure below?

A. 17.2

B. 16

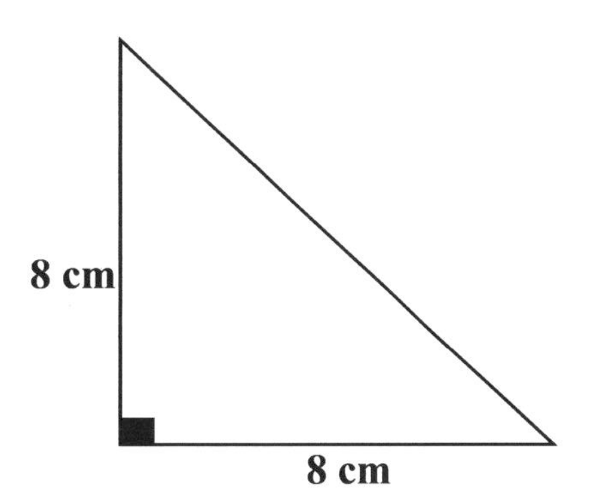

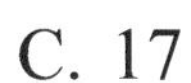
C. 17

D. 27.2

11) What is the range of the following set of data: $2, -2, 6, -1, 1, 6, 4, 3, 9$?

A. 2

B. 9

C. 6

D. 11

12) Alex starts a saving account with $60. Every week he adds $7 to his account.

Which equation can be used to determine the number of weeks w, after which

Alex's accounts reaches $185?

A. $7w + 185 = 60$

B. $7 + w = 185$

C. $7w + 60 = 185$

D. $7w - 60 = 185$

13) The medals won by United States, Australia and Spain during a basketball competition are shown in the table below:

United States	Australia	Spain
6	9	9

Out of the medals won by these three countries, what percentage of medals did the United States win?

A. 10%

B. 25%

C. 15%

D. 75%

14) A girl in State A spent $56 before a 6.75% sales tax and a girl in State B spent $52 before an 6.25% sales tax. How much more money did the girl from State A spend than the girl from State B after sales tax was applied? Round to the nearest hundredth.

A. 4.53

B. 45.30

C. 42

D. 15.43

15) A school has 465 students and 22 chemistry teachers and 15 physics teachers. What is the ratio between the number of physics teachers and the number of students at the school?

A. $\frac{1}{31}$

B. $\frac{9}{31}$

C. $\frac{1}{37}$

D. $\frac{15}{22}$

16) James has his own lawn mowing service. The maximum James charges to mow a lawn is $42. Which inequality represents the amount James could charge, P, to mow a lawn?

A. $P < 42$

B. $P = 42$

C. $P \leq 42$

D. $P \geq 42$

17) What is the value of this expression $19 \div 0.76$?

A. 0.75

B. 4.25

C. 20

D. 25

18) The ratio of boys to girls in Maria Club is the same as the ratio of boys to girls in Hudson Club. There are 32 boys and 56 girls in Maria Club. There are 12 boys in Hudson Club. How many girls are in Hudson Club?

A. 28

B. 7

C. 18

D. 21

19) On average, Simone drinks $\frac{2}{5}$ of a 6-ounce glass of coffee in $\frac{2}{3}$ hour. How much coffee does she drink in an hour?

A. 0.7 ounces

B. 1.6 ounces

C. 3.6 ounces

D. 6.3 ounces

20) Line P, R, and S intersect each other, as shown in below diagram. Based on the angle measures, what is the value of θ?

A. 36°

B. 108°

C. 72°

D. 132°

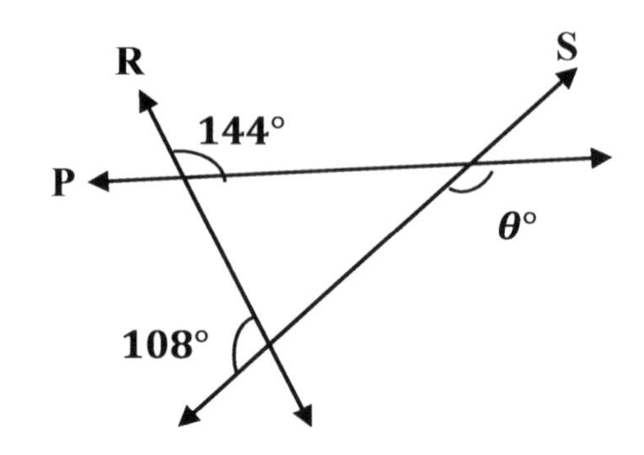

21) Which expression is represented by the model below?

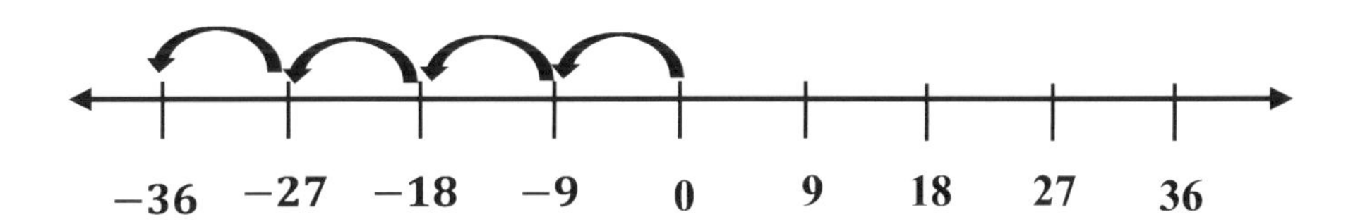

A. $-4 . (-9)$

B. $(-4) . 9$

C. $4 . (-9)$

D. $4 . 9$

22) The table below shows the distance, y, a lion can travel in mile in x hour.

Time (x, hour)	Distance (y, mile)
7	364
14	728
21	1,092
28	1,456
35	1,820

Based on the information in the table, which equation can be used to model the relationship between x and y?

A. $y = x + 7$

B. $y = 7x$

C. $y = x + 364$

D. $y = 52x$

23) Mia has a loan of $42,750. This loan has a simple interest rate of 2.6% per year. What is the amount of interest that Mia will be charged on this loan at the end of one year?

A. $22,22.25

B. $11,115

C. $18,111

D. $1,111.5

24) The spinner shown has eight congruent sections.

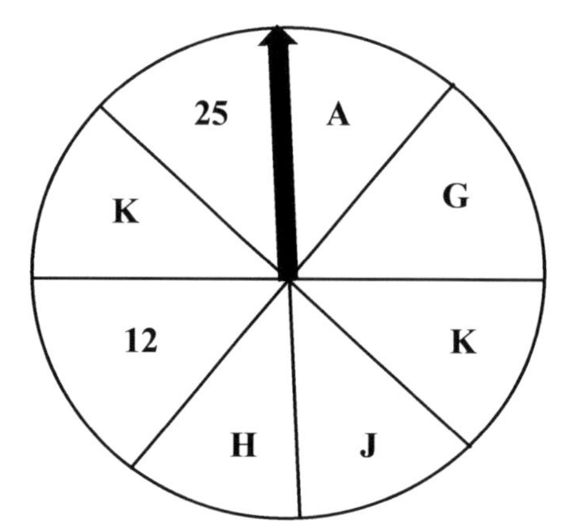

The spinner is spun 160 times. What is a reasonable prediction for the number of times the spinner will land on a letter?

A. 40

B. 12

C. 110

D. 120

25) Which graph best represents the distance a car travels when going 20 miles per hour?

A.

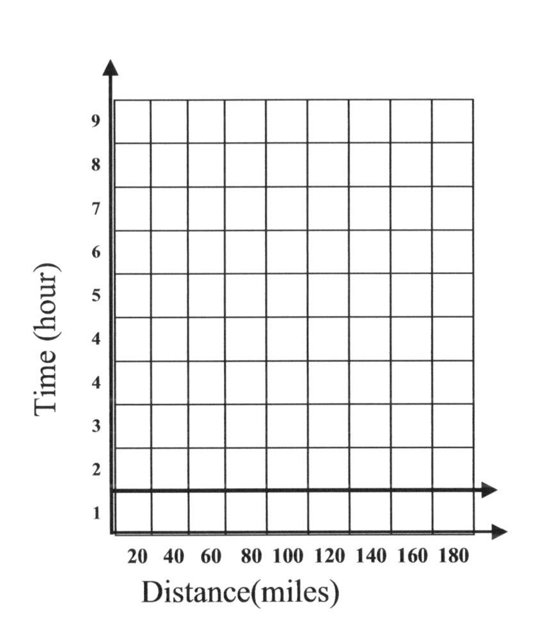

B.

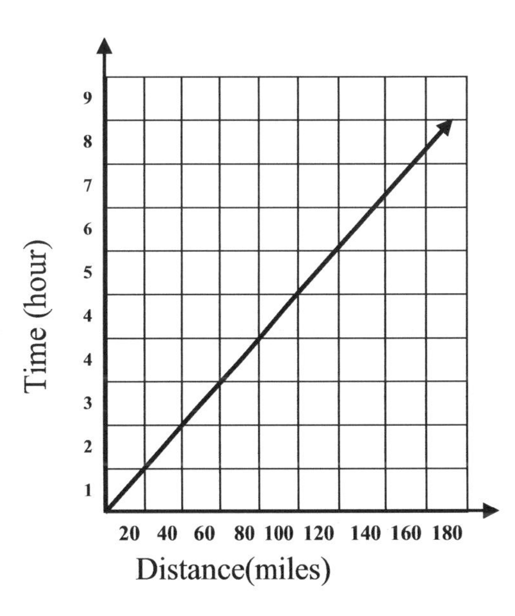

C.

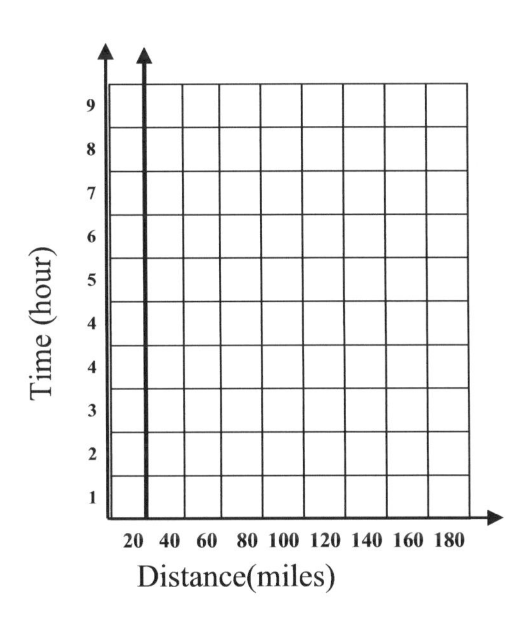

D.

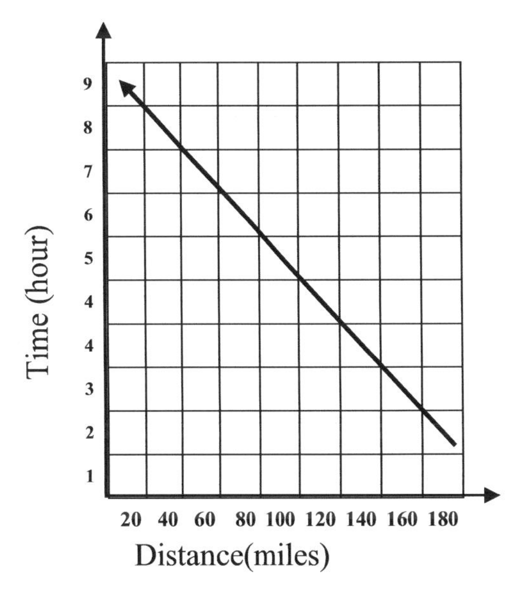

26) The temperature is shown in the table below, on each of day in the week for a city in February. What is the mean temperature, in the city for that week?

A. -17

B. -19.5

C. -8

D. -7.2

Day	Temperature (°F)
Monday	-26
Tuesday	-33
Wednesday	-17
Thursday	-9
Friday	0
Saturday	11
Sunday	18

27) Which arithmetic sequence is represented by the expression $5m - 2$, where m represents the position of a term in the sequence?

A. 8, 13, 19, 24, 29, …

B. 8, 13, 18, 23, 28, …

C. 13, 18, 23, 27, 32, …

D. 13, 17, 19, 23, 28, …

28) Which expression is equivalent to $-28 - 350d$?

A. $-14(2 - 35d)$

B. $-7(40d - 7)$

C. $-14(2 + 25d)$

D. $-225\ d$

29) The dot plots show how many minutes per day do 7th grade study math after school at two different schools on one day.

Number of minuets study in school 1

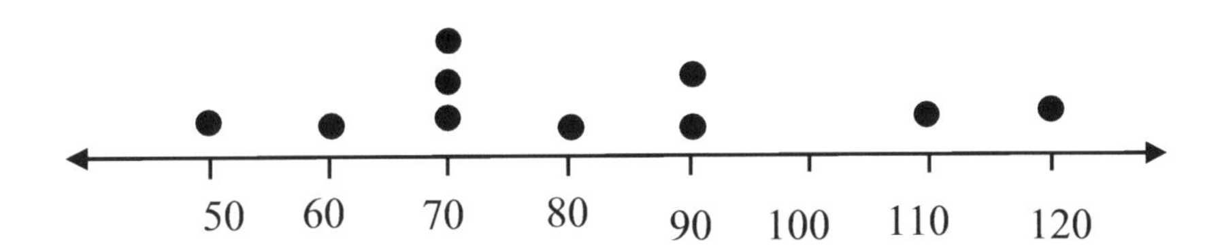

Number of minuets study in school 2

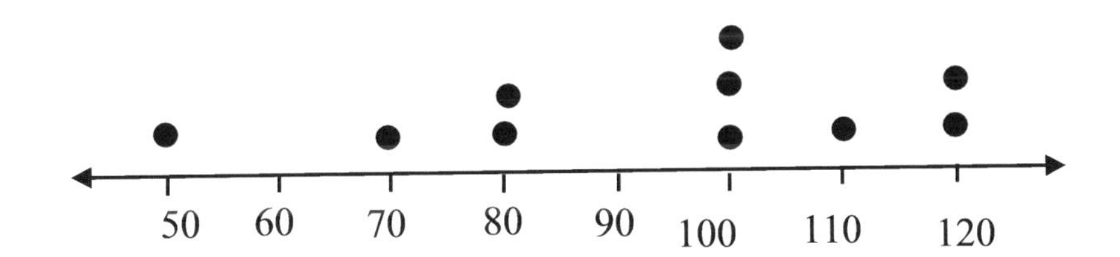

Which statement is supported by the information in the dot plots?

A. The mode of the data for School 2 is greater than the mode of the data for School 1.

B. The mean of the data for School 1 is greater than the mean of the data for School 2.

C. The median of the data for School 2 is smaller than the median of the data for School 1.

D. The median and mean of the data for two schools are equal.

30) Which number represents the probability of an event that is unlikely to occur?

A. 0.98

B. 1.5

C. 0.20

D. 0.52

Smarter Balanced Assessment Consortium

SBAC Practice Test 4

Mathematics

GRADE 7

Administered *Month Year*

1) What is the decimal equivalent of the fraction $\frac{47}{22}$?

A. 2.16

B. $2.\overline{136}$

C. $2.1\overline{36}$

D. 2.13

2) Thomas is shareholder of a company. The price of stock is $86.95 on the morning of day 1. Thomas records the change in the price of the stock in the chart below at the end of each day, but some information is missing.

Day	Change in Price ($)
1	+ 0.58
2	+0.76
3	
4	−0.64
5	

The change in the price for day 3 is $\frac{3}{8}$ of the change in the price for day 4. At the end of day 5, the price of Thomas's stock is $88.29. What is the change, in dollars, in the price of the stock for day 5?

A. −0.12

B. 0.88

C. −0.88

D. 0.12

3) Kevin adds $\frac{3}{7}$ cups of sugar into a mixture every $\frac{1}{4}$ hour. What is the rate, in cups per minute, at which Kevin adds sugar to the mixture?

A. $\frac{1}{35}$

B. $2\frac{1}{7}$

C. $\frac{1}{17}$

D. $\frac{1}{105}$

4) A box of ball contains 6 blue balls, 8 red balls, 6 black balls, and 2 green balls. All the balls are the same size and shape. Brian will select a ball at random. Which of the following best describes the probability that Brian will select a green ball?

A. unlikely

B. certain

C. likely

D. impossible

5) The first number in a pattern is 5. Each following number is found by subtracting 6 from the previous number. What is the seventh number in the pattern?

A. -36

B. -21

C. -19

D. -31

6) Evelyn opened a bank account. She adds the same amount of money to her account each month. The table below shows the amounts of money in her account at the ends of certain numbers of months.

How much money does Evelyn add to her bank account each month?

Month	Amount
3	$54
6	$108
9	$162

A. $9

B. $15

C. $18

D. $36

7) Using data from house sales, probabilities for the story of a house sold were calculated. The probabilities for two story are listed below.

- The probability a house sold has one story is 0.42.
- The probability a house sold has two story 0.52.

Based on these probabilities, how many of the next 500 houses sold are likely to be one story and how many are likely to be two story?

A. one story: 42, two story: 52

B. one story: 105, two story: 130

C. one story: 210, two story: 260

D. one story: 260, two story: 210

8) Mr. Turner is digging a trench to put in the new school sprinkler system. Every $\frac{1}{6}$ hour, the length of his trench increases by $\frac{3}{5}$ foot. By how much does the length, in feet, of Mr. Turner's trench increase each hour?

A. $\frac{1}{5}$

B. $\frac{3}{10}$

C. $\frac{5}{18}$

D. $\frac{18}{5}$

9) Multiply: $3\frac{5}{9} \times \frac{-5}{9}$

A. $-1\frac{7}{81}$

B. $-1\frac{5}{9}$

C. $-1\frac{79}{81}$

D. -3

10) Brendan charges $26 per hour plus $70 to enter data. He accepted a project for no more than $730. Which inequality can be used to determine all the possible numbers of hours (x) it took the man to enter the data?

A. $26x + 70 \leq 730$

B. $26x + 70 > 730$

C. $70x + 26 < 730$

D. $70x + 26 \geq 730$

11) Use the coordinate grid below to answer the question. What is the circumference of the circle?

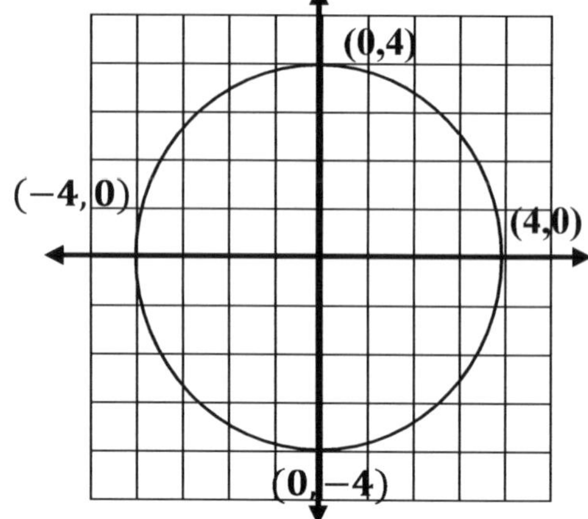

A. 12.56

B. 25.12

C. 36.24

D. 169.12

12) The temperature is 7° F. As a cold front move in, the temperature drops 5° F per half hour. What is the temperature at the end of 2 hours?

A. 13°F

B. 20°F

C. −20°F

D. −13°F

13) A printer originally cost h dollars, including tax. Eddy purchased the printer when it was on sale for 26% off its original cost. Which of the following expressions represents the final cost, in dollars, of the printer Eddy purchased?

A. $h + 0.74$

B. $h - 0.26$

C. $0.74h$

D. $0.26h$

14) Use the set of data below. What is the median of the list of numbers?

38, 23, 35, 28, 23, 30

A. 35

B. 29

C. 28

D. 30

15) Asher worked out at a gym for 4 hours. His workout consisted of jogging for 58 minutes, playing volleyball for 62 minutes, and playing billiards for the remaining amount of time. What percentage of Asher's workout was spent playing billiards?

A. 50%

B. 45%

C. 15%

D. 55%

16) The angle measures of a triangle GBD are shown in the diagram. What is the value of $\angle B$?

A. 19

B. 72

C. 95

D. 118

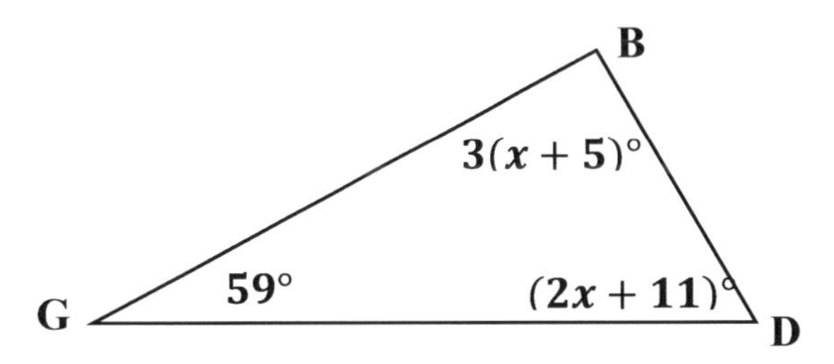

17) Triangle PRS is shown on the grid below.

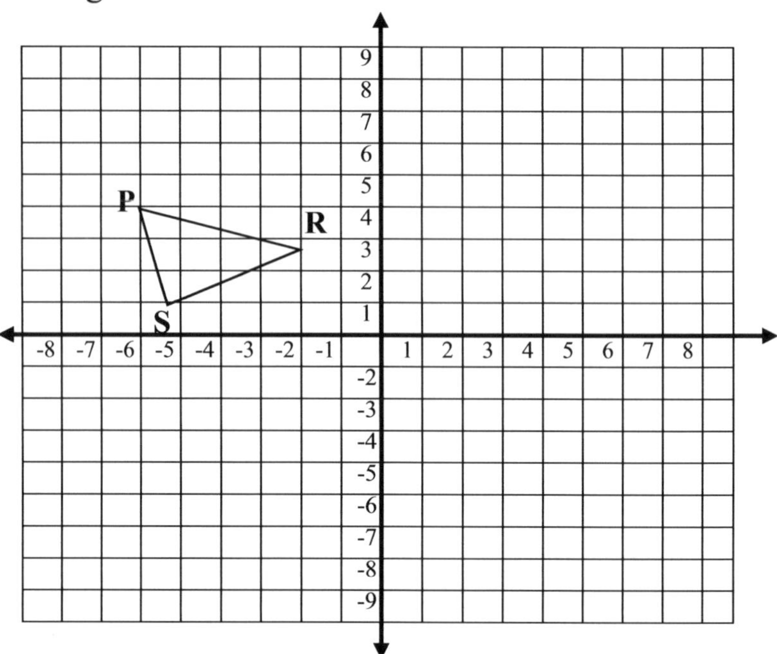

If triangle PRS is reflected across the y-axis to form triangle P′R′S′, which ordered pair represents the coordinates of P′?

A. $(-6, -4)$

B. $(6, -4)$

C. $(6,4)$

D. $(4, -6)$

18) What is the solution set for the inequality $-7x + 30 > -12$?

A. $x > 6$

B. $x < 6$

C. $x > -5$

D. $x < -5$

19) In a party people drink 118.28 liters of juices. There are approximately 29.57 milliliters in 1 fluid ounce. Which measurement is closest to the number of fluid ounces in 118.28 liters?

A. 0.004 fl oz

B. 2,782.48 fl oz

C. 2,008.84 fl oz

D. 4,000 fl oz

20) The dimensions of a square pyramid are shown in the diagram. What is the volume of the square pyramid in cubic inches?

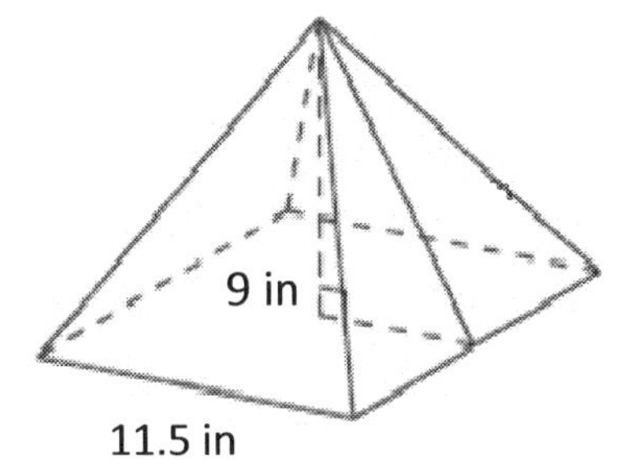

A. 396.75 in^3

B. 396 in^3

C. 693.5 in^3

D. 961.75 in^3

21) Water is poured to fill a pool in the shape of a rectangular prism. The pool is 15 feet long, 6.8 feet wide, and 12.7 feet high. How much cubic feet of water are needed to fill the pool?

A. 1,295.4 ft^3

B. 129.45 ft^3

C. 129.54 ft^3

D. 129.94 ft^3

22) The store manager spent $13,720 to buy a new freezer and 28 tables. The total purchase is represented by this equation, where v stands for the value of each table purchased: $28v + 1{,}120 = 13{,}720$

What was the cost of each table that the manager purchased?

A. $505

B. $500

C. $544

D. $450

23) In a city, at 2:15 A.M., the temperature was -6°F. At 2:15 P.M., the temperature was 15°F. Which expression represents the increase in temperature?

A. $-6 - 15$

B. $|-6 - 15|$

C. $|-6| - 15$

D. $-6 - |15|$

24) Angles α and β are complementary angles. Angles α and are supplementary angles. The degree measure of angle β is $110°$. What is the measure of angle γ?

A. $20°$

B. $110°$

C. $70°$

D. $60°$

25) The bar graph shows a company's income and expenses over the last 5 years.

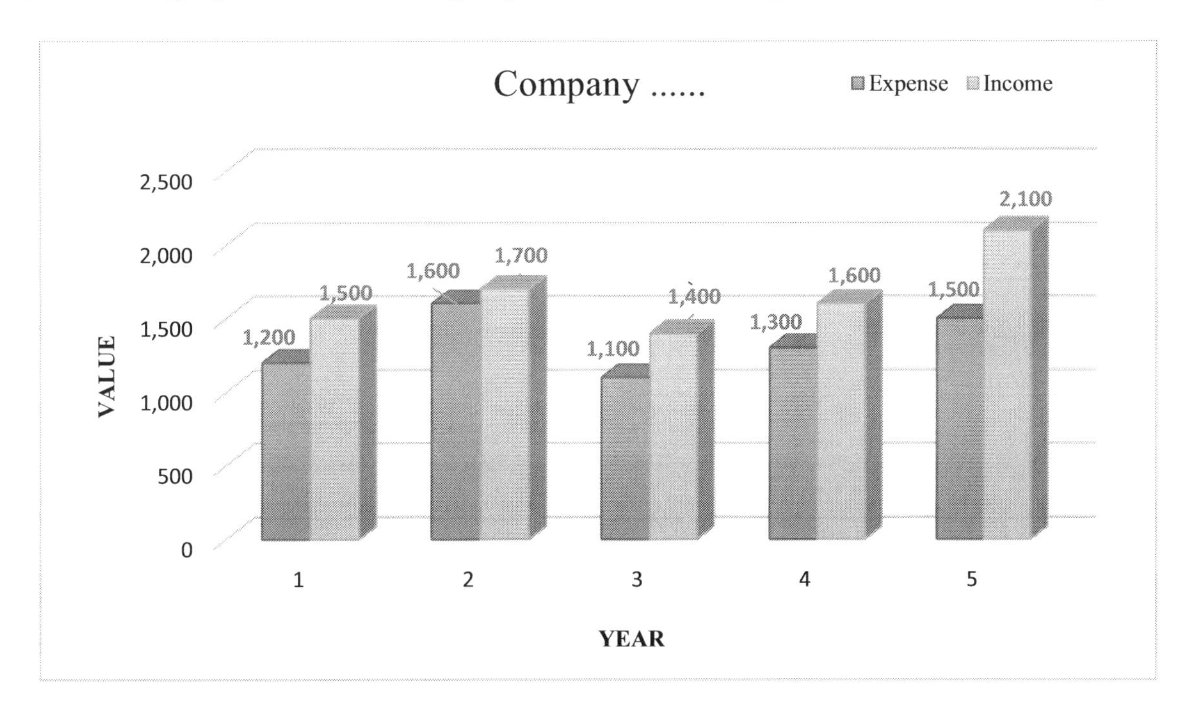

Which statement is supported by the information in the graph?

A. Expenses have increased \$400 each year over the last 5 years.

B. The income in Year 5 was 15% more than the income in Year 1.

C. The combined income in Years 3, 4, and 5 was equal to the combined expenses in Years 2, 3, and 4.

D. Expenses in the year 3 was more than half of the income in the year 4.

26) Which expression is equivalent to the $(3n - 9) - \frac{1}{3}(8 - 9n) + \frac{5}{3}$?

A. -6

B. $-3n - 10$

C. $6n - 10$

D. $6n - 6$

27) Patricia bought a bottle of 16-ounce balsamic vinegar for \$13.06. She used 35% of the balsamic vinegar in two weeks. Which of the following is closest to the cost of the balsamic she used?

A. \$0.45

B. \$7.47

C. \$4.57

D. \$6.75

28) A scale drawing of triangle DEF that will be used on a wall is shown below. What is the perimeter, in meter, of the actual triangle used on the wall?

Scale: 1 cm: $3\frac{1}{5}$ m

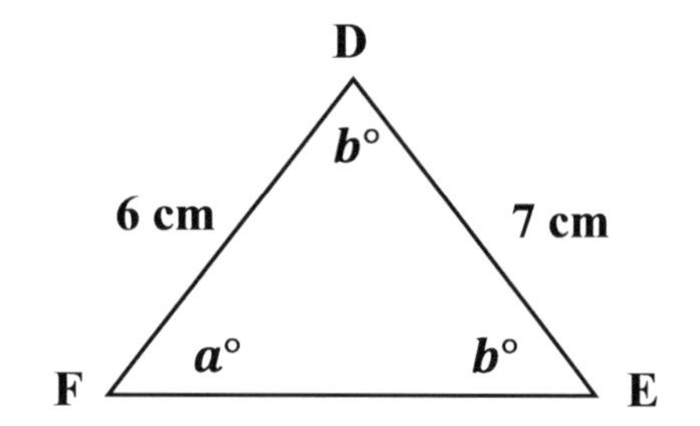

A. $21\frac{2}{5}$

B. $57\frac{3}{5}$

C. $67\frac{1}{5}$

D. $47\frac{1}{5}$

29) The ratio of boys to girls in Geometry class is 4 to 5. There are 25 girls in the class. What is the total number of students in Geometry class?

A. 65

B. 45

C. 20

D. 15

30) A group of employees have their weight recorded to make a data set. The mean, median, mode, and range of the data set are recorded. Then, the weight of the manager is included to make a new data set. The manager's weight is more than all but one of the employees. Which measure must be the same when the manager's weight included?

A. Mean

B. Mode

C. Median

D. Range

Smarter Balanced Assessment Consortium

SBAC Practice Test 5

Mathematics

GRADE 7

Administered *Month Year*

1) Peter paid for 5 sandwiches.

- Each sandwich cost 12.95.
- He paid for 4 bags of fries that each cost $2.45.

Which equation can be used to determine the total amount, y, Peter paid?

A. $y = 5(12.95) + 4(2.45)x$

B. $y = (12.95 + 2.45)x$

C. $y = 5(12.95) + 4(2.45)$

D. $y = 12.95x + 4(2.45)$

2) What is the decimal equivalent of the fraction $\frac{10}{27}$?

A. 0.37

B. $0.3\overline{70}$

C. $0.\overline{370}$

D. 0.370

3) The circumference of a circle is 14 π centimeters. What is the area of the circle in terms of π?

A. 14 π

B. 49 π

C. 196 π

D. 28 π

4) What is the volume of rectangular prism when the two triangular prisms below are stuck together?

A. 78 in^3

B. 39 in^3

C. 156 in^3

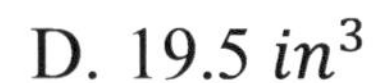
D. 19.5 in^3

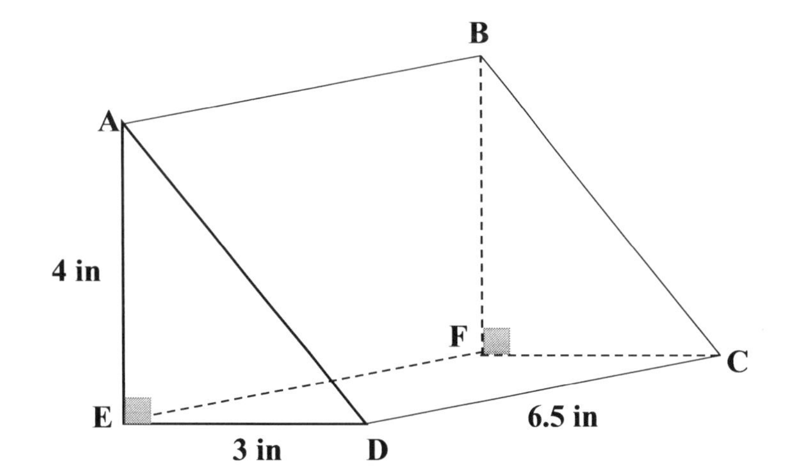

5) Which number line Shows the solution to the inequality $-4x - 1 < -7$?

A.

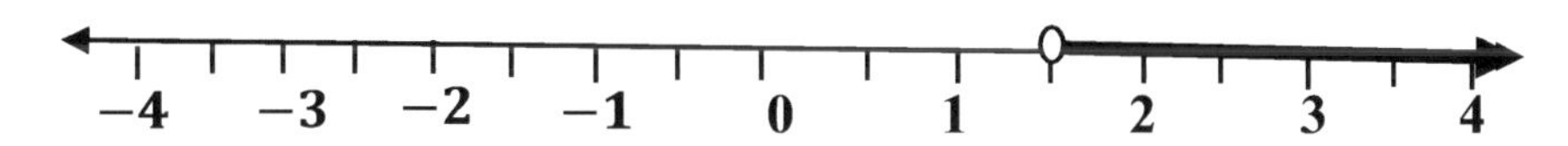

B.

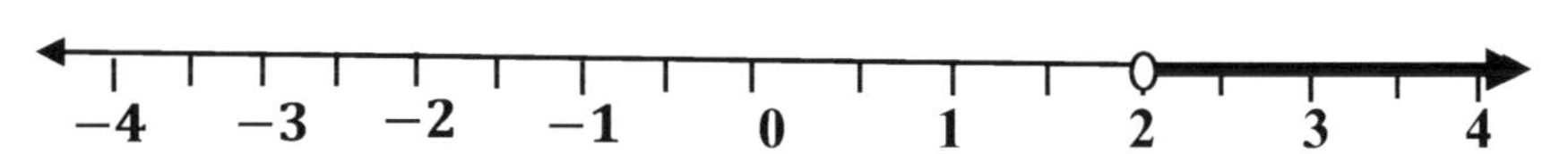

C.

D.

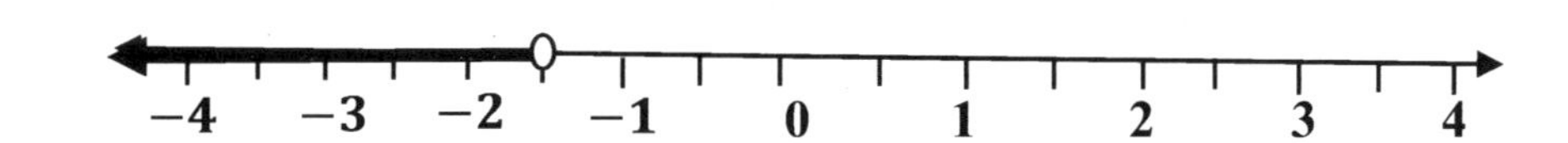

6) What is the value of $(4 + 7)^2 + (4 - 7)^2$?

A. -33

B. 64

C. 112

D. 130

7) Arsan has $8 to spend on school supplies. The following table shows the price of each item in the school store. No sale tax is charged on these items.

Which the combination of items can Arsan buy with his $8?

A. 3 Notebooks and 3 Pens

B. 5 Folders and 3 Erasers

C. 3 Notebooks and 3 Folders

D. 5 Erasers and 3 Pens.

Item	Price
Notebook	$1.85
Pen	$0.90
Eraser	$ 0.76
Folder	$1.18

8) If 16% of x is 64, what is 23% of x?

A. 92

B. 14.72

C. 10.24

D. 2.36

9) If all variables are positive, find the square root of $\frac{9x^7y^2}{49x}$?

A. $\frac{3}{7}x^4y$

B. $\frac{7y}{3x^3}$

C. $\frac{3}{7}x^3y$

D. $5\frac{4}{9}x^2y^3$

10) Which is closest to the perimeter of the right triangle in the figure below?

A. 15.6

B. 16.6

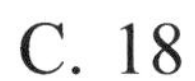
C. 18

D. 20.4

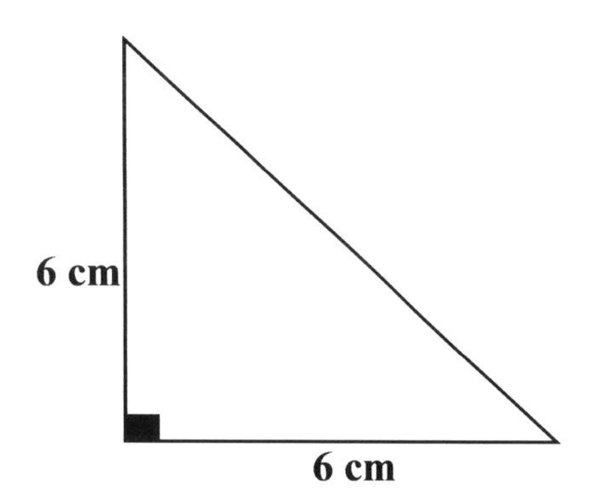

11) What is the range of the following set of data: $1, -1, 5, -2, 3, 6, 5, 2, 8$?

A. 5

B. 6

C. 12

D. 10

12) Alex starts a saving account with \$80. Every week he adds \$9 to his account.

Which equation can be used to determine the number of weeks w, after which

Alex's accounts reaches \$215?

A. $9w + 215 = 80$

B. $9 + w = 215$

C. $9w + 80 = 215$

D. $9w - 80 = 215$

13) The medals won by United States, Australia and Spain during a basketball competition are shown in the table below:

United States	Australia	Spain
8	5	7

Out of the medals won by these three countries, what percentage of medals did the United States win?

A. 8%

B. 40%

C. 35%

D. 80%

14) A girl in State A spent $64 before a 7.25% sales tax and a girl in State B spent $48 before an 8.25% sales tax. How much more money did the girl from State A spend than the girl from State B after sales tax was applied? Round to the nearest hundredth.

A. 16.68

B. 22.80

C. 15

D. 16

15) A school has 450 students and 32 chemistry teachers and 18 physics teachers. What is the ratio between the number of physics teachers and the number of students at the school?

A. $\frac{1}{25}$

B. $\frac{9}{16}$

C. $\frac{1}{9}$

D. $\frac{16}{225}$

16) James has his own lawn mowing service. The maximum James charges to mow a lawn is $30. Which inequality represents the amount James could charge, P, to mow a lawn?

A. $P < 30$

B. $P = 30$

C. $P \geq 30$

D. $P \leq 30$

17) What is the value of this expression $16 \div 0.64$?

A. 0.04

B. 0.25

C. 4

D. 25

18) The ratio of boys to girls in Maria Club is the same as the ratio of boys to girls in Hudson Club. There are 24 boys and 40 girls in Maria Club. There are 18 boys in Hudson Club. How many girls are in Hudson Club?

A. 31

B. 9

C. 15

D. 25

19) On average, Simone drinks $\frac{3}{5}$ of a 4-ounce glass of coffee in $\frac{3}{4}$ hour. How much coffee does she drink in an hour?

A. 0.75 ounces

B. 1.8 ounces

C. 3.2 ounces

D. 6.4 ounces

20) Line P, R, and S intersect each other, as shown in below diagram. Based on the angle measures, what is the value of θ?

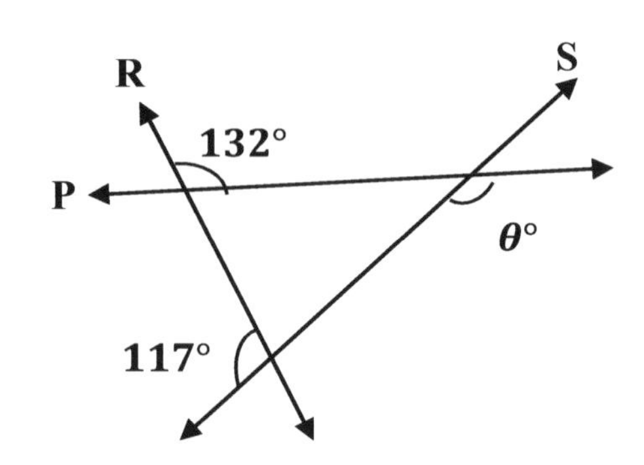

A. 69°

B. 111°

C. 63°

D. 132°

21) Which expression is represented by the model below?

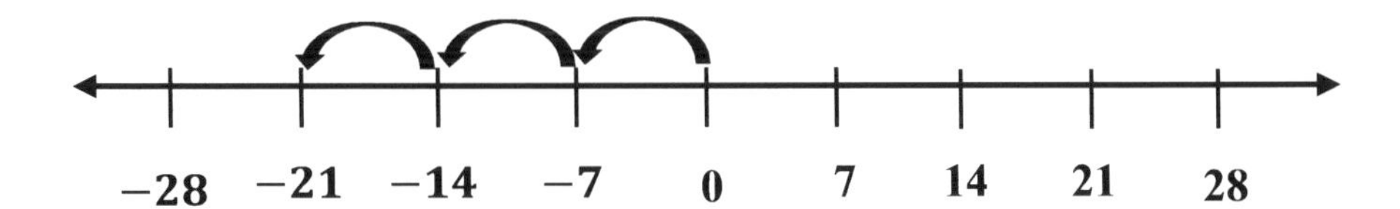

A. $-3 . (-7)$

B. $(-3) . 7$

C. $3 . (-7)$

D. $3 . 7$

22) The table below shows the distance, y, a lion can travel in mile in x hour.

Time (x, hour)	Distance (y, mile)
6	282
12	564
18	846
24	1,128
30	1,410

Based on the information in the table, which equation can be used to model the relationship between x and y?

A. $y = x + 6$

B. $y = 6x$

C. $y = x + 282$

D. $y = 47x$

23) Mia has a loan of $36,850. This loan has a simple interest rate of 3.5% per year. What is the amount of interest that Mia will be charged on this loan at the end of one year?

A. $38,139.75

B. $12,284

C. $18,232

D. $1,289.75

24) The spinner shown has eight congruent sections.

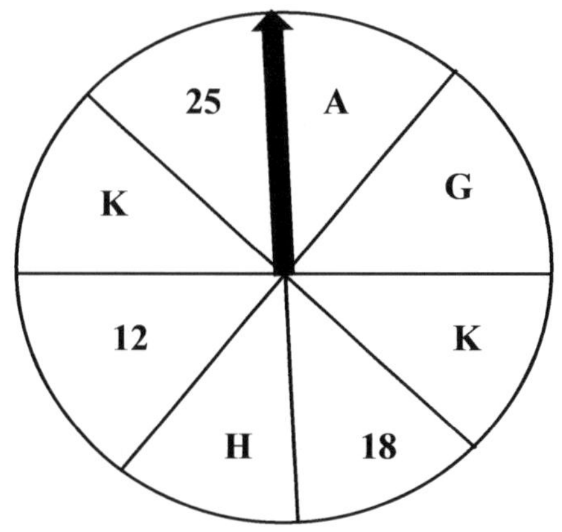

The spinner is spun 120 times. What is a reasonable prediction for the number of times the spinner will land on a letter?

A. 3

B. 15

C. 45

D. 75

25) Which graph best represents the distance a car travels when going 25 miles per hour?

A.

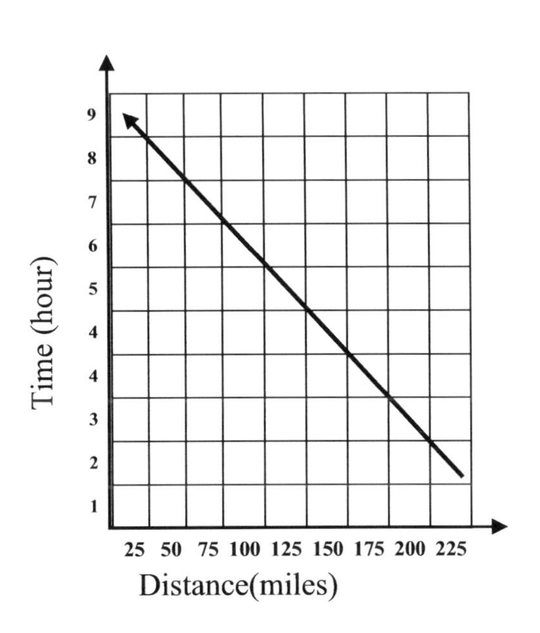

B.

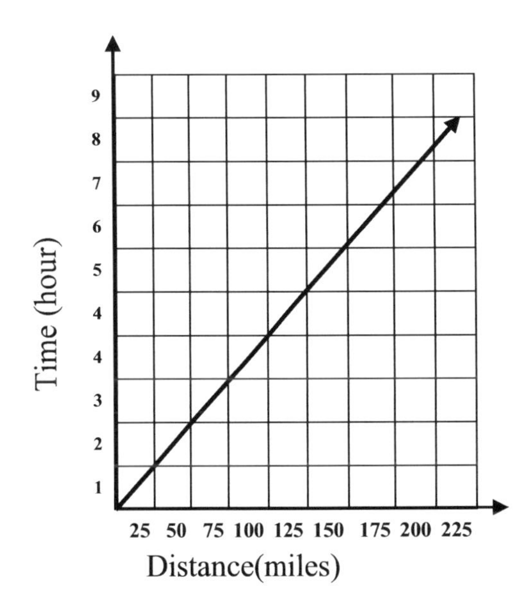

C.

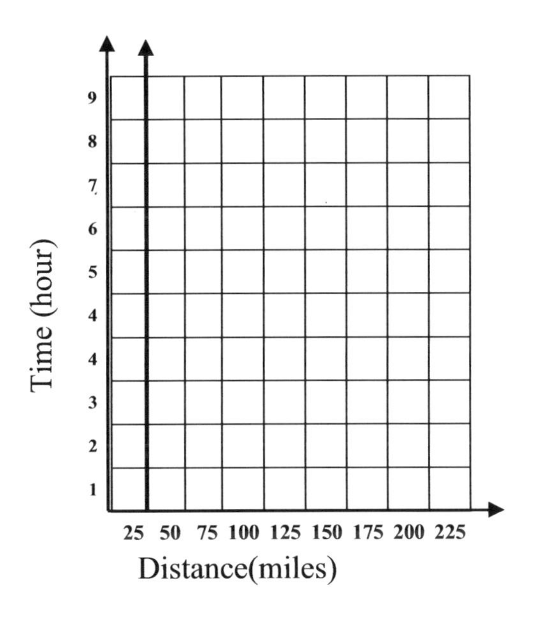

D.

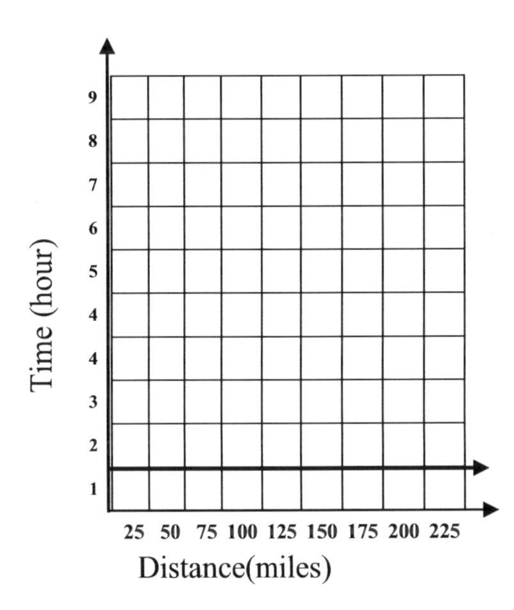

26) The temperature is shown in the table below, on each of day in the week for a city in February. What is the mean temperature, in the city for that week?

A. -15

B. -17.5

C. -7

D. -8.2

Day	Temperature (°F)
Monday	-25
Tuesday	-31
Wednesday	-15
Thursday	0
Friday	12
Saturday	-6
Sunday	16

27) Which arithmetic sequence is represented by the expression $4m - 1$, where m represents the position of a term in the sequence?

A. 12, 16, 20, 24, 28, …

B. 11, 15, 19, 23, 27, …

C. 12, 11, 10, 9, 8, …

D. 7, 8, 9, 10, 11, …

28) Which expression is equivalent to $-24 - 420d$?

A. $-12(2 - 35d)$

B. $-6(70d - 4)$

C. $-12(2 + 35d)$

D. $-264\,d$

29) The dot plots show how many minutes per day do 7th grade study math after school at two different schools on one day.

Number of minuets study in school 1

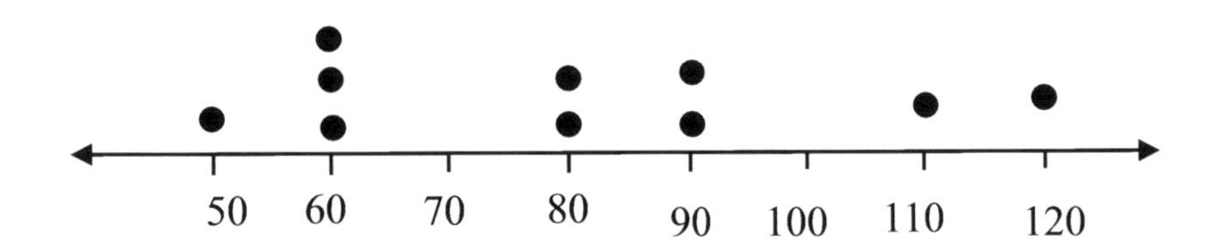

Number of minuets study in school 2

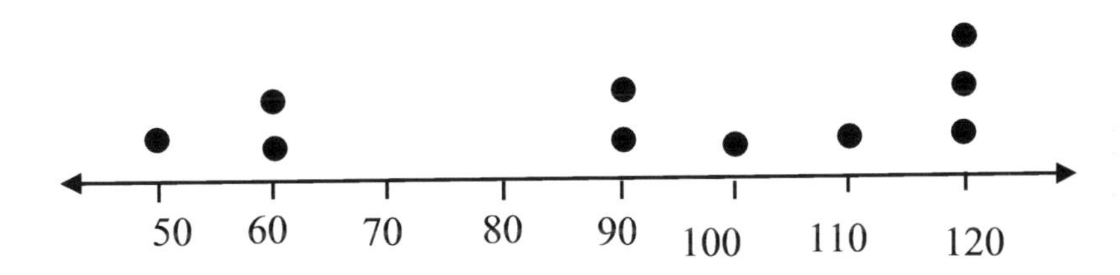

Which statement is supported by the information in the dot plots?

A. The mode of the data for School 2 is greater than the mode of the data for School 1.

B. The mean of the data for School 1 is greater than the mean of the data for School 2.

C. The median of the data for School 2 is smaller than the median of the data for School 1.

D. The median and mean of the data for two schools are equal.

30) Which number represents the probability of an event that is very likely to occur?

A. 0.12

B. 1.3

C. 0.89

D. 0.09

Smarter Balanced Assessment Consortium

SBAC Practice Test 6

Mathematics

GRADE 7

Administered *Month Year*

1) What is the decimal equivalent of the fraction $\frac{37}{33}$?

A. 1.12

B. $1.1\overline{21}$

C. $1.\overline{12}$

D. 1.121

2) Thomas is shareholder of a company. The price of stock is $98.67 on the morning of day 1. Thomas records the change in the price of the stock in the chart below at the end of each day, but some information is missing.

Day	Change in Price ($)
1	+ 0.84
2	+0.93
3	
4	−0.72
5	

The change in the price for day 3 is $\frac{2}{3}$ of the change in the price for day 4. At the end of day 5, the price of Thomas's stock is $100.05. What is the change, in dollars, in the price of the stock for day 5?

A. −0.15

B. 0.81

C. 0.33

D. 1.38

3) Kevin adds $\frac{2}{5}$ cups of sugar into a mixture every $\frac{1}{6}$ hour. What is the rate, in cups per minute, at which Kevin adds sugar to the mixture?

A. $\frac{1}{25}$

B. $2\frac{1}{5}$

C. $\frac{1}{50}$

D. $\frac{1}{15}$

4) A box of ball contains 5 blue balls, 2 red balls, 7 black balls, and 1 green balls. All the balls are the same size and shape. Brian will select a ball at random. Which of the following best describes the probability that Brian will select a green ball?

A. likely

B. certain

C. unlikely

D. impossible

5) The first number in a pattern is 7. Each following number is found by subtracting 8 from the previous number. What is the sixth number in the pattern?

A. -55

B. -41

C. -29

D. -33

6) Evelyn opened a bank account. She adds the same amount of money to her account each month. The table below shows the amounts of money in her account at the ends of certain numbers of months.

How much money does Evelyn add to her bank account each month?

Month	Amount
4	$64
8	$128
12	$192

A. $8

B. $12

C. $16

D. $32

7) Using data from house sales, probabilities for the story of a house sold were calculated. The probabilities for two story are listed below.

- The probability a house sold has one story is 0.34.
- The probability a house sold has two story 0.46.

Based on these probabilities, how many of the next 300 houses sold are likely to be one story and how many are likely to be two story?

A. one story: 34, two story: 46

B. one story: 66, two story: 54

C. one story: 102, two story: 138

D. one story: 240, two story: 240

8) Mr. Turner is digging a trench to put in the new school sprinkler system. Every $\frac{1}{4}$ hour, the length of his trench increases by $\frac{2}{3}$ foot. By how much does the length, in feet, of Mr. Turner's trench increase each hour?

A. $\frac{1}{6}$

B. $\frac{3}{7}$

C. $\frac{11}{12}$

D. $\frac{8}{3}$

9) Multiply: $2\frac{7}{11} \times \frac{-7}{11}$

A. $6\frac{5}{11}$

B. $-18\frac{5}{11}$

C. $-1\frac{82}{121}$

D. 2

10) Brendan charges \$35 per hour plus \$50 to enter data. He accepted a project for no more than \$640. Which inequality can be used to determine all the possible numbers of hours (x) it took the man to enter the data?

A. $35x + 50 \leq 640$

B. $35x + 50 > 640$

C. $50x + 35 < 640$

D. $50x + 35 \geq 640$

11) Use the coordinate grid below to answer the question. What is the circumference of the circle?

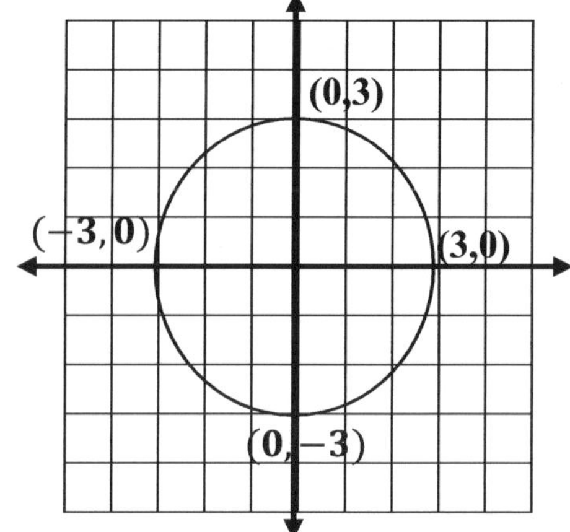

A. 9.42

B. 18.84

C. 37.68

D. 188.4

12) The temperature is 9° F. As a cold front move in, the temperature drops 4° F per half hour. What is the temperature at the end of 2 hours?

A. 7°F

B. 25°F

C. −25°F

D. −7°F

13) A printer originally cost h dollars, including tax. Eddy purchased the printer when it was on sale for 38% off its original cost. Which of the following expressions represents the final cost, in dollars, of the printer Eddy purchased?

A. $h + 0.62$

B. $h - 0.38$

C. $0.62h$

D. $0.38h$

14) Use the set of data below. What is the median of the list of numbers?

35, 20, 32, 25, 20, 27

A. 32

B. 26

C. 25

D. 20

15) Asher worked out at a gym for 3 hours. His workout consisted of jogging for 52 minutes, playing volleyball for 65 minutes, and playing billiards for the remaining amount of time. What percentage of Asher's workout was spent playing billiards?

A. 35%

B. 63%

C. 17%

D. 65%

16) The angle measures of a triangle GBD are shown in the diagram. What is the value of $\angle B$?

A. 33.5

B. 72

C. 23

D. 63

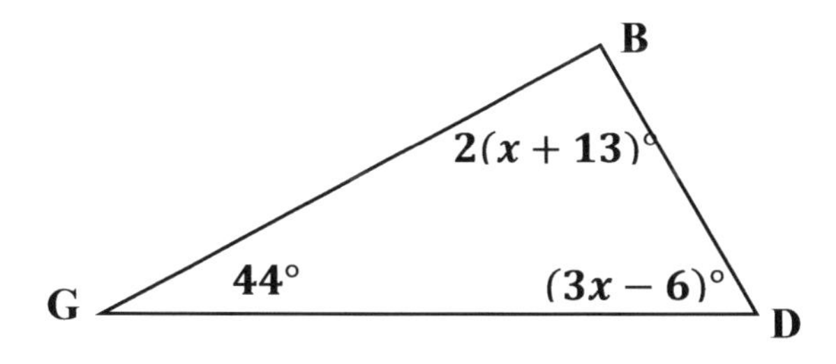

17) Triangle PRS is shown on the grid below.

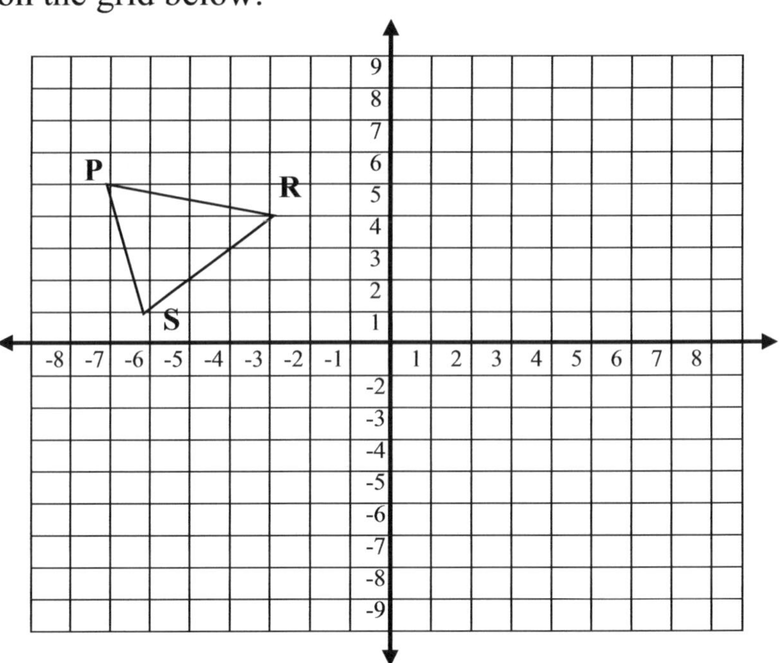

If triangle PRS is reflected across the y-axis to form triangle P′R′S′, which ordered pair represents the coordinates of P′?

A. $(-7,-5)$

B. $(7,-5)$

C. $(7,5)$

D. $(5,-7)$

18) What is the solution set for the inequality $-8x + 40 > -16$?

A. $x > 7$

B. $x < 7$

C. $x > -3$

D. $x < -3$

19) In a party people drink 59.15 liters of juices. There are approximately 29.57 milliliters in 1 fluid ounce. Which measurement is closest to the number of fluid ounces in 59.15 liters?

A. 0.002 fl. oz

B. 1,749.24 fl. oz

C. 1,095.72 fl. oz

D. 2,000 fl. oz

20) The dimensions of a square pyramid are shown in the diagram. What is the volume of the square pyramid in cubic inches?

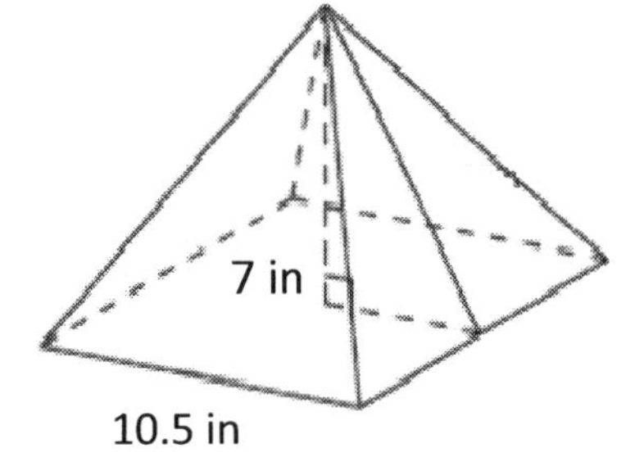

A. 257.25 in^3

B. 294 in^3

C. 514.5 in^3

D. 771.75 in^3

21) Water is poured to fill a pool in the shape of a rectangular prism. The pool is 17 feet long, 8.4 feet wide, and 11.25 feet high. How much cubic feet of water are needed to fill the pool?

A. 1,606.5 ft^3

B. 571.25 ft^3

C. 537.75 ft^3

D. 109.95 ft^3

22) The store manager spent $12,850 to buy a new freezer and 25 tables. The total purchase is represented by this equation, where v stands for the value of each table purchased: $25v + 1{,}350 = 12{,}850$

What was the cost of each table that the manager purchased?

A. $568

B. $540

C. $514

D. $460

23) In a city, at 1:30 A.M., the temperature was -9°F. At 1:30 P.M., the temperature was 17°F. Which expression represents the increase in temperature?

A. $-9 - 17$

B. $|-9 - 17|$

C. $|-9| - 17$

D. $-9 - |17|$

24) Angles α and β are complementary angles. Angles α and γ are supplementary angles. The degree measure of angle β is 90°. What is the measure of angle γ?

A. 0°

B. 45°

C. 90°

D. 180°

25) The bar graph shows a company's income and expenses over the last 5 years.

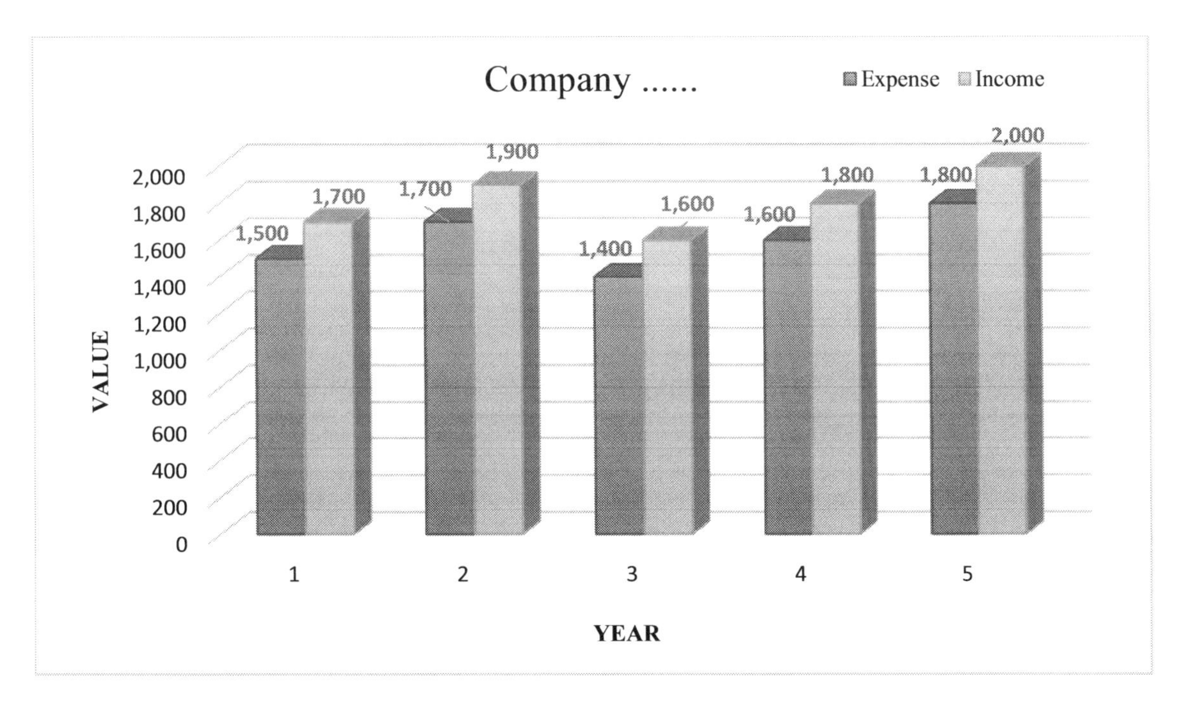

Which statement is supported by the information in the graph?

A. Expenses have increased \$200 each year over the last 5 years.

B. The income in Year 5 was 20 % more than the income in Year 1.

C. The combined income in Years 3, 4, and 5 was equal to the combined expenses in Years 2, 3, and 4.

D. Expenses in the year 3 was more than half of the income in the year 4.

26) Which expression is equivalent to the $(2n - 8) - \frac{1}{2}(9 - 4n) + \frac{5}{2}$?

A. -10

B. $-2n - 10$

C. $4n - 10$

D. $4n - 6$

27) Patricia bought a bottle of 16-ounce balsamic vinegar for $11.08. She used 45% of the balsamic vinegar in two weeks. Which of the following is closest to the cost of the balsamic she used?

A. $0.31

B. $2.49

C. $4.99

D. $6.09

28) A scale drawing of triangle DEF that will be used on a wall is shown below. What is the perimeter, in meter, of the actual triangle used on the wall?

Scale: 1 cm: $2\frac{1}{4}$ m

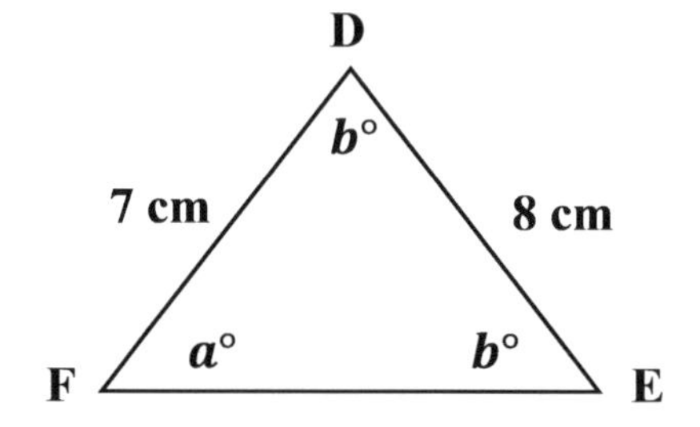

A. $28\frac{3}{4}$

B. $51\frac{3}{4}$

C. $49\frac{1}{2}$

D. $47\frac{1}{4}$

29) The ratio of boys to girls in Geometry class is 2 to 3. There are 18 girls in the class. What is the total number of students in Geometry class?

A. 12

B. 30

C. 45

D. 27

30) A group of employees have their weight recorded to make a data set. The mean, median, mode, and range of the data set are recorded. Then, the weight of the manager is included to make a new data set. The manager's weight is more than all but one of the employees. Which measure must be the same when the manager's weight included?

A. Mean

B. Median

C. Mode

D. Range

Answers and Explanations

Answer Key

Now, it's time to review your results to see where you went wrong and what areas you need to improve!

SBAC Math Practice Tests

Practice Test 1

1	C	11	D	21	C
2	C	12	C	22	D
3	B	13	B	23	D
4	A	14	A	24	D
5	A	15	A	25	B
6	D	16	C	26	C
7	B	17	D	27	B
8	A	18	D	28	C
9	C	19	C	29	A
10	D	20	B	30	C

Practice Test 2

1	C	11	B	21	A
2	B	12	D	22	D
3	A	13	C	23	B
4	A	14	B	24	A
5	D	15	A	25	D
6	C	16	B	26	C
7	C	17	C	27	C
8	D	18	B	28	C
9	C	19	D	29	B
10	A	20	A	30	D

SBAC Math Practice Tests

Practice Test 3

1	C	11	D	21	C
2	C	12	C	22	D
3	B	13	B	23	D
4	A	14	A	24	D
5	A	15	A	25	B
6	D	16	C	26	C
7	C	17	D	27	B
8	A	18	D	28	C
9	C	19	C	29	A
10	D	20	B	30	C

Practice Test 4

1	C	11	B	21	A
2	B	12	D	22	D
3	A	13	C	23	B
4	A	14	B	24	A
5	D	15	A	25	D
6	C	16	B	26	C
7	C	17	C	27	C
8	D	18	B	28	C
9	C	19	D	29	B
10	A	20	A	30	D

SBAC Math Practice Tests

Practice Test 5

1	C	11	D	21	C
2	C	12	C	22	D
3	B	13	B	23	D
4	A	14	A	24	D
5	A	15	A	25	B
6	D	16	D	26	C
7	C	17	D	27	B
8	A	18	D	28	C
9	C	19	C	29	A
10	D	20	B	30	C

Practice Test 6

1	C	11	B	21	A
2	B	12	D	22	D
3	A	13	C	23	B
4	C	14	B	24	A
5	D	15	A	25	D
6	C	16	B	26	C
7	C	17	C	27	C
8	D	18	B	28	C
9	C	19	D	29	B
10	A	20	A	30	D

Practice Test 1

Answers and Explanations

1) Answer: C

Let y be the total amount paid.

We have been given that Peter bought 9 sandwiches that each cost the $12.34.

So, the cost of 9 sandwiches would be 9(12.34). He paid for 8 bags of fries that each cost the $2.25. So, the cost of 8 bags would be 8(2.25)

Then, the total cost of sandwiches and fries would be $y = 9(12.34) + 8(2.25)$

2) Answer: C

To get the answer to 7 over 22 as a decimal, we divide 7 by 22.

$\frac{7}{22} = 0.3181818 \ldots. = 0.3\overline{18}$

3) Answer: B

Use the formula of circumference of circles.

Circumference = $\pi d = 2\pi\,(r) = 16\,\pi \Rightarrow r = 8$

Radius of the circle is 8. Now, use the areas formula:

Area = $\pi r^2 \Rightarrow$ Area = $\pi(8)^2 \Rightarrow$ Area = $64\,\pi$

4) Answer: A

The volume of a triangular prism is the base times the height. $V = Bh$

Area of the base = $\frac{1}{2} b.h \rightarrow B = \frac{1}{2} \times 6 \times 8 = 24$

$V = B.h = 24 \times 9.5 = 228$; we need two triangular prisms, then $2 \times 228 = 456$

5) Answer: A

$-5x - 4 < -9$, add 4 to both sides $-5x < -5$ divide each term by -5

If an inequality is multiplied or divided by a negative number, you must change the direction of the inequality, then $x > 1$

6) Answer: D

$(2+7)^2+(2-7)^2=(9)^2+(-5)^2=81+25=106$

7) Answer: B

A. $(6\times 3.75)+(4\times 0.90)=22.50+3.60=26.10>16$

B. $(4\times 1.95)+(6\times 1.30)=7.80+7.80=15.60<16$

C. $(5\times 3.75)+(2\times 1.95)=18.75+3.90=22.65>16$

D. $(7\times 1.30)+(9\times 0.90)=9.10+8.10=17.2>16$

8) Answer: A

$0.17\times x=51\rightarrow x=\frac{51}{0.17}=\frac{5{,}100}{17}=300$

$45\% of\ 300=0.45\times 300=135$

9) Answer: C

$\sqrt{\frac{25x^9y^5}{49xy}}=\sqrt{\frac{25}{49}\times\frac{x^9y^5}{xy}}=\sqrt{\frac{25}{49}x^8y^4}=\frac{5}{7}x^4y^2$

10) Answer: D

Use Pythagorean Theorem: $a^2+b^2=c^2$

$6^2+6^2=c^2\Rightarrow 36+36=c^2\Rightarrow 72=c^2\Rightarrow c=\sqrt{72=36\times 2}=6\sqrt{2}=$

$6\times 1.4=8.5$

Perimeter of triangle $=a+b+c=6+6+8.5=20.5$

11) Answer: D

Range= Largest Value – Smallest Value → R = $12-(-3)=15$

12) Answer: C

To start, assign the variables to unknowns, known values to constants, and relate them by the relations between the variables and constants. Then,

$9w+80=192$

13) Answer: B

Use percent formula: part = $\frac{\text{percent}}{100}\times$ whole

Whole = $5 + 8 + 7 = 20$

$5 = \frac{\text{percent}}{100} \times 20 \Rightarrow 5 = \frac{\text{percent} \times 20}{100} \Rightarrow 500 = \text{percent} \times 20 \Rightarrow \text{percent} = \frac{500}{20} = 25,$

Therefore United States win 25% of medals.

14) Answer: A

Multiply the price by the sales tax to find out how much money the sales tax will add, then Add the original price and the sales tax.

State A: $44 \times 0.025 = 1.1$

$44 + 1.1 = 45.1$

State B: $40 \times 0.0175 = 0.7$

$40 + 0.7 = 40.7$

Then take the difference: $45.1 - 40.7 = 4.4$

15) Answer: A

16 physics teachers to 432 students are 16:432, 1:27

16) Answer: C

At least and Minimum – means greater than or equal to

At most, no more than, and Maximum – means less than or equal to

More than – means greater than

Less than – means less than

Then, $P \leq 51$

17) Answer: D

$\frac{22}{0.55}$ Multiply the numerator and denominator by 100: $\frac{2{,}200}{55} = 40$

18) Answer: D

The ratio of boys to girls in Maria Club: 24:42 = 4:7

The ratio of boys to girls in Hudson Club: 4:7

4 :7 same as 20: 35.

So, there are 35 girls in Hudson club.

19) Answer: C

$\frac{3}{4} \times 5 = 3.75$

$\frac{1}{4} \times 60 = 15\, min$

$\frac{3.75}{15} = \frac{x}{60} \rightarrow 15x = 3.75 \times 60 \rightarrow x = \frac{225}{15} = 15$ Ounces

20) Answer: B

Supplementary angles are two angles that have a sum of $180°$

$ine\ R: 180° - 151° = 29°, Line\ S: 180° - 102° = 78°$

then in the triangle: $180° - (78° + 29°) = 73°$

$line\ P: \theta° = 180° - 73° = 107°$

21) Answer: C

$3 \times (-8) = -32$

22) Answer: D

$\frac{228}{6} = 38$

$\frac{456-228}{12-6} = \frac{228}{6} = 38$

23) Answer: D

$38{,}450 \times 0.018 = 692.1$

24) Answer: D

The diagram shows a spinner with 8 sections. The probability land on letter is $\frac{5}{8}$

The spinner is spun 240 times, then the prediction is: $240 \times \frac{5}{8} = 150$ times.

25) Answer: B

A linear equation is a relationship between two variables, and application of linear equations can be found in distance problems.

$d = rt$ or distance equals rate (speed) times time.

$d = 1 \times 30 = 30$,then $(1,30), (2,60), (3,90)(4,120), \ldots$

26) Answer: C

average (mean) $= \frac{sum of terms}{number of terms} = \frac{(-24)+(-31)+(-15)+(-11)+0+12+20}{7} = \frac{-49}{7} =$

$-7°F$

27) Answer: B

$m = 1 \rightarrow 6m - 5 = 6(1) - 5 = 1$

$m = 2 \rightarrow 6m - 5 = 6(2) - 5 = 7$

$m = 3 \rightarrow 6m - 5 = 6(3) - 5 = 13$

$m = 4 \rightarrow 6m - 5 = 6(4) - 5 = 19$

$m = 5 \rightarrow 6m - 5 = 6(5) - 5 = 25$

28) Answer: C

$-26 - 312d = -13(2 + 24d)$

29) Answer: A

Let's find the mode, mean (average), and median of the number of minutes for each school.

Number of Minutes for school 1: 50, 50, 60, 70, 70, 70, 80, 90, 110, 110

Mean(average) $= \frac{sum of terms}{number of terms} = \frac{50+50+60+70+70+70+80+90+110+110}{10} = \frac{760}{10} = 76$

Median is the number in the middle. Since there are an even number of items in the resulting list, the median is the average of the two middle numbers.

Median of the data is $(70 + 70) \div 2 = 70$

Mode is the number which appears most often in a set of numbers. Therefore, there is no mode in the set of numbers. Mode is 70.

Number of Minutes for school 2: 50, 60, 80, 90, 100, 100, 110, 110, 110, 120

Mean$= \frac{50+60+80+90+100+100+110+110+110+120}{10} = \frac{930}{10} = 93$

Median: $(100 + 100) \div 2 = 100$

Mode: 110

30) Answer: C

We often describe the probability of something happening with words like impossible, unlikely, as likely as unlikely, equally likely, likely, and certain. The probability of an event occurring is represented by a ratio. A ratio is a number that is between 0 and 1 and can include 0 and 1. An event is impossible if it has a probability of 0. An event is certain if it has the probability of 1

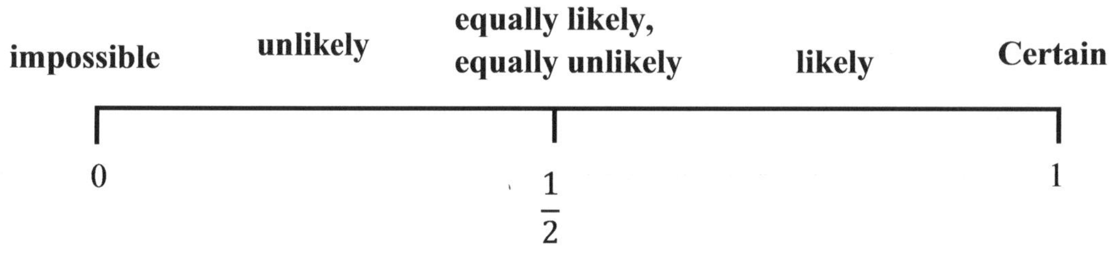

impossible	unlikely	equally likely, equally unlikely	likely	Certain
0		$\frac{1}{2}$		1

Practice Test 2

Answers and Explanations

1) Answer: C

Divided 34 by 11: $\frac{34}{11} = 3.09090909 \ldots. = 3.\overline{09}$

2) Answer: B

Day 3: $-0.70 \times \frac{2}{5} = -0.28$

Change price in days: $\big(75.85 + 0.52 + 0.66 + (-0.28) + (-0.70)\big) = 76.05$

Day 5: $76.92 - 76.05 = 0.87$

3) Answer: A

$1hour = 60\ min \rightarrow \frac{1}{5} \times 60 = 12\ min$

Rate: $\frac{\frac{2}{9}}{12} = \frac{x}{1} \rightarrow 12x = \frac{2}{9} \rightarrow x = \frac{2}{108} = \frac{1}{54}$

4) Answer: A

Probability $= \frac{\text{number of desired outcomes}}{\text{number of total outcomes}} = \frac{3}{11+9+10+3} = \frac{3}{33} = \frac{1}{11}$

If an event has a 0 probability this means that it can never happen

If an event has a 1 probability it will certainly happen

If an event has a 0.5 probability, it has an equal chance of happening or not happening.

If an event has a probability between 0 and 0.5, then it is unlikely to happen.

If an event has a probability between 0.5 and 1, then it is likely to happen.

5) Answer: D

The pattern is: $4, -1, -6, -11, -16, -21$

6) Answer: C

$64 \div 4 = 16$

$112 \div 7 = 16$

7) Answer: C

likely one-story house is: $0.35 \times 400 = 140$

likely of two-story house is: $0.50 \times 400 = 200$

8) Answer: D

Write the ratio and solve for x.

$\frac{1}{\frac{1}{9}} = \frac{x}{\frac{2}{7}}$ (Cross multiply) $\Rightarrow \frac{1}{9}x = \frac{2}{7} \Rightarrow x = \frac{18}{7}$

9) Answer: C

$4\frac{3}{8} \times \frac{-3}{8} = \frac{35}{8} \times \frac{-3}{8} = -\frac{105}{64} = -1\frac{41}{64}$

10) Answer: A

Hour: $x \rightarrow 31$ per hour: $31x$

Plus: add $(+)$, no more than: $\leq$; Then, $31x + 60 \leq 820$

11) Answer: B

By grid line: $d = 4$

Or distance for two points $(0,4), (0,-4)$: $d = \sqrt{(0-0)^2 + (2-(-2))^2} = 4$

$d = 4 \rightarrow$ Circumference $= \pi d = \pi(4) = 4\pi = 12.56$

12) Answer: D

3 hours equal 6 half hours

$6 \times 4° = 24° \rightarrow 8 - 24 = -16°\text{F}$

13) Answer: C

If the price of a printer is decreased by 32% then: $100\% - 32\% = 68\%$

$68\ \%\ of\ \text{h} = 0.68 \times \text{h} = 0.68\text{h}$

14) Answer: B

The median of a set of data is the value located in the middle of the data set. To find median, first list numbers in order from smallest to largest:

21, 21, 26, 28, 36, 42

Since there are an even number of items in the resulting list, the median is the average of the two middle numbers.

Median= $(26 + 28) \div 2 = 27$

15) Answer: A

Each hour is 60 minutes, so we have $3 \times 60 = 180$ minutes of workout time for Asher. We subtract off the jogging and playing volleyball time to get the time Asher playing billiards: $180 - 42 - 48 = 90$ minutes.

$$percent = \left(\frac{\text{part}}{\text{whole}}\right) \times 100 \rightarrow percent = \left(\frac{90}{180}\right) \times 100 \rightarrow percent = 50\%$$

16) Answer: B

$4(x + 6) + (x + 10) + 46 = 180 \rightarrow 4x + 24 + x + 10 + 46 = 180$

$5x + 80 = 180 \rightarrow 5x = 100 \rightarrow x = \frac{100}{5} = 20 \Rightarrow x = 20°$

$\angle B = 4(x + 6°) = 4(20° + 6°) = 104°$

17) Answer: C

The reflection of the point (x, y) across the y-axis is the point $(-x, y)$.

If you reflect a point across the y −axis, the y −coordinate is the same, but the x −coordinate is changed into its opposite.

Reflection of $(-6, -2) \rightarrow (6, -2)$

18) Answer: B

$-5x + 12 > -8$ (Subtract 12 from both sides)

$-5x > -20$ (Divide both side by -5, remember negative change the sign)

$x < 4$

19) Answer: D

1 L=1,000 mL

1 fl oz = 29.57 mL→ 1 fl oz =0.02957 L

$$88.71 \text{ L} = \frac{88.71}{0.02957} \times \frac{100{,}000}{100{,}000} = \frac{8{,}871{,}000}{2{,}957} = 3{,}000\, fl\ oz$$

20) Answer: A

Volume of a square-based pyramid: V $=\frac{1}{3}B.h$, where V is the volume and B is the area of the base. Then, V $=\frac{1}{3}\times 8.5\times 8.5\times 6=\frac{433.5}{3}=144.5$

21) Answer: A

Use formula of rectangle prism volume.

V = (length) (width) (height) ⇒ V $=12\times 7.5\times 11.8=1{,}062\ ft^3$

22) Answer: D

$25v + 2{,}340 = 11{,}840$ (subtract 2,340 from both sides)

$\rightarrow 25v = 9{,}500$ (divide both sides by 25) $\rightarrow v = \$380$ cost of each table.

23) Answer: B

Difference of temperature is: $|t_2-t_1|=|17-(-4)|=|17+4|=|-4-17|$

24) Answer: A

Supplementary angles are two angles with a sum of 180 degrees.

$\alpha+\beta=180°$ and $\beta=118° \Longrightarrow \alpha=180°-118°=62°$

complementary angles are two angles with a sum of 90 degrees.

$\alpha+\gamma=90$ and $\alpha=62° \Longrightarrow \gamma=90°-62°=28°$

25) Answer: D

A. Expenses in Year 2: 2,000 $\rightarrow 2{,}000+500=2{,}500\neq$ Year 3

B. Income in Year 1: 1,900 and $20\%\ of\ 1{,}900=380\rightarrow 1{,}900+380=$

$2{,}280\neq$ income of Year 5

C. Income's year 3, 4, and 5: $1{,}800+2{,}000+2{,}500=6{,}300$

Expense's year 2, 3, and 4: $2{,}000+1{,}500+1{,}700=5{,}200\neq 6{,}300$

D. Half of Year incomes 4: $\frac{2{,}000}{2}=1{,}000<1{,}500$ Expenses in Year 3

26) Answer: C

$(4n-7)-\frac{1}{2}(11-8n)+\frac{7}{2}=4n-7-\frac{11}{2}+4n+\frac{7}{2}=8n-9$

27) Answer: C

If you ever need to find the percentage of something you just times it by the fraction. So, all you need to do to figure this out is to find 45% of 12.14 which is approximately 5.46.

28) Answer: C

$\angle D = \angle E = b° \rightarrow FD = FE = 5cm$

Perimeter of triangle: $5 + 5 + 6 = 16\ cm$

Actual triangle: $16 \times 2\frac{1}{3} = 16 \times \frac{7}{3} = \frac{112}{3} = 37\frac{1}{3}\ cm$

29) Answer: B

The ratio of boy to girls is 2:7. Therefore, there are 7 girls out of 9 students. To find the answer, first divide the number of girls by 7, then multiply the result by 9.

$42 \div 7 = 6 \Rightarrow 6 \times 9 = 54$

30) Answer: D

A. does not consider that the manager's weight could be different from the mean weight and would change the value of the mean.

B. does not consider that the manager could be the same weight as the employee and that this weight could be the new mode.

C. does not consider that the nth heaviest employee (in the middle) could be different weight and adding a weight that is greater than either of these weights to the data set would change the value of the median.

D. Corrects.

Practice Test 3

Answers and Explanations

1) Answer: C

Let y be the total amount paid.

We have been given that Peter bought 7 sandwiches that each cost the $10.22. So, the cost of 7 sandwiches would be 7(10.22). He paid for 6 bags of fries that each cost the $1.87. So, the cost of 6 bags would be 6(1.87)

Then, the total cost of sandwiches and fries would be $y = 7(10.22) + 6(1.87)$

2) Answer: C

To get the answer to 6 over 11 as a decimal, we divide 6 by 11.

$\frac{6}{11} = 0.545454 \ldots. = 0.\overline{54}$

3) Answer: B

Use the formula of circumference of circles.

Circumference = $\pi d = 2\pi\,(r) = 18\,\pi \Rightarrow r = 9$

Radius of the circle is 9. Now, use the areas formula:

Area = $\pi r^2 \Rightarrow$ Area = $\pi(9)^2 \Rightarrow$ Area = $81\,\pi$

4) Answer: A

The volume of a triangular prism is the base times the height. $V = Bh$

Area of the base = $\frac{1}{2} b.h \rightarrow B = \frac{1}{2} \times 4 \times 5 = 10$

$V = B.h = 10 \times 7.5 = 75$ and we need two triangular prisms, then $2 \times 75 = 150$

5) Answer: A

$-7x - 3 < -10$, add 3 to both sides $-7x < -7$ divide each term by -7

If an inequality is multiplied or divided by a negative number, you must change the direction of the inequality, then $x > 1$

6) Answer: D

$(3+8)^2+(3-8)^2=(11)^2+(-5)^2=121+25=146$

7) Answer: B

A. $(4\times 2.25)+(2\times 1.10)=9+2.20=11.2>11$

B. $(3\times 2.15)+(5\times 0.85)=6.45+4.25=10.7<11$

C. $(2\times 2.25)+(4\times 2.15)=4.5+8.6=13.1>11$

D. $(6\times 0.85)+(6\times 1.10)=5.1+6.6=11.7>11$

8) Answer: A

$0.18\times x=72\rightarrow x=\frac{72}{0.18}=\frac{7{,}200}{18}=400$

$35\% of\ 400=0.35\times 400=140$

9) Answer: C

$\sqrt{\frac{16x^9y^3}{81xy}}=\sqrt{\frac{16}{81}\times\frac{x^9y^3}{xy}}=\sqrt{\frac{16}{81}x^8y^2}=\frac{4}{9}x^4y$

10) Answer: D

Use Pythagorean Theorem: $a^2+b^2=c^2$

$8^2+8^2=c^2\Rightarrow 64+64=c^2\Rightarrow 128=c^2\Rightarrow c=\sqrt{128}=8\sqrt{2}=8\times 1.4=11.2$

Perimeter of triangle $=a+b+c=8+8+11.2=27.2$

11) Answer: D

Range= Largest Value – Smallest Value $\rightarrow$ R $=9-(-2)=11$

12) Answer: C

To start, assign the variables to unknowns, known values to constants, and relate them by the relations between the variables and constants. Then

$7w+60=185$

13) Answer: B

Use percent formula: part $=\frac{\text{percent}}{100}\times$ whole

Whole = $6 + 9 + 9 = 24$

$6 = \frac{\text{percent}}{100} \times 24 \Rightarrow 6 = \frac{\text{percent} \times 24}{100} \Rightarrow 600 = \text{percent} \times 24 \Rightarrow \text{percent} = \frac{600}{24} = 25$,

Therefore United States win 25% of medals.

14) Answer: A

Multiply the price by the sales tax to find out how much money the sales tax will add, then Add the original price and the sales tax.

State A: $56 \times 0.0675 = 3.78$

$56 + 3.78 = 59.78$

State B: $52 \times 0.0625 = 3.25$

$52 + 3.25 = 55.25$

Then take the difference: $59.78 - 55.25 = 4.53$

15) Answer: A

15 physics teachers to 465 students is 15:465, 1:31

16) Answer: C

At least and Minimum – means greater than or equal to

At most, no more than, and Maximum – means less than or equal to

More than – means greater than

Less than – means less than

Then, $P \leq 42$

17) Answer: D

$\frac{19}{0.76}$ Multiply the numerator and denominator by 100: $\frac{1{,}900}{76} = 25$

18) Answer: D

The ratio of boys to girls in Maria Club: 32:56 = 4:7

The ratio of boys to girls in Hudson Club: 4:7

4 :7 same as 12: 21.

So, there are 21 girls in Hudson club.

19) Answer: C

$\frac{2}{5} \times 6 = 2.4$

$\frac{2}{3} \times 60 = 40\ min$

$\frac{2.4}{40} = \frac{x}{60} \rightarrow 40x = 2.4 \times 60 \rightarrow x = \frac{144}{40} = 3.6$ Ounces

20) Answer: B

Supplementary angles are two angles that have a sum of 180°

$ine\ R$: $180° - 144° = 36°$, $Line\ S$: $180° - 108° = 72°$

then in the triangle: $180° - (72° + 36°) = 72°$

$line\ P$: $\theta° = 180° - 72° = 108°$

21) Answer: C

$4 \times (-9) = -36$

22) Answer: D

$\frac{364}{7} = 52$

$\frac{728-364}{14-7} = \frac{364}{7} = 52$

23) Answer: D

$42{,}750 \times 0.026 = 1{,}111.5$

24) Answer: D

The diagram shows a spinner with 8 sections. The probability land on letter is $\frac{6}{8}$

The spinner is spun 160 times, then the prediction is: $160 \times \frac{6}{8} = 120$ times.

25) Answer: B

A linear equation is a relationship between two variables, and application of linear equations can be found in distance problems.

$d = rt$ or distance equals rate (speed) times time.

$d = 1 \times 20 = 20$,then $(1,20), (2,40), (3,60)(4,80), \ldots$

26) Answer: C

average (mean) $= \frac{sum of terms}{number of terms} = \frac{(-26)+(-33)+(-17)+(-9)+0+11+18}{7} = \frac{-56}{7} = -8°F$

27) Answer: B

$m = 1 \rightarrow 5m - 2 = 5(1) - 2 = 3$

$m = 2 \rightarrow 5m - 2 = 5(2) - 2 = 8$

$m = 3 \rightarrow 5m - 2 = 5(3) - 2 = 13$

$m = 4 \rightarrow 5m - 2 = 5(4) - 2 = 18$

$m = 5 \rightarrow 5m - 2 = 5(5) - 2 = 23$

28) Answer: C

$-28 - 350d = -14(2 + 25d)$

29) Answer: A

Let's find the mode, mean (average), and median of the number of minutes for each school.

Number of Minutes for school 1: 50, 60, 70, 70, 70, 80, 90, 90, 110, 120

Mean(average) $= \frac{sum of terms}{number of terms} = \frac{50+60+70+70+70+80+90+90+110+120}{10} = \frac{810}{10} = 81$

Median is the number in the middle. Since there are an even number of items in the resulting list, the median is the average of the two middle numbers.

Median of the data is $(70 + 80) \div 2 = 75$

Mode is the number which appears most often in a set of numbers. Therefore, there is no mode in the set of numbers. Mode is:70

Number of Minutes for school 2: 50, 70, 80, 80, 100, 100, 100, 110, 120, 120

Mean$= \frac{50+70+80+80+100+100+100+110+120+120}{10} = \frac{930}{10} = 93$

Median: $(100 + 100) \div 2 = 100$

Mode: 100

30) Answer: C

We often describe the probability of something happening with words like impossible, unlikely, as likely as unlikely, equally likely, likely, and certain. The probability of an event occurring is represented by a ratio. A ratio is a number that is between 0 and 1 and can include 0 and 1. An event is impossible if it has a probability of 0. An event is certain if it has the probability of 1.

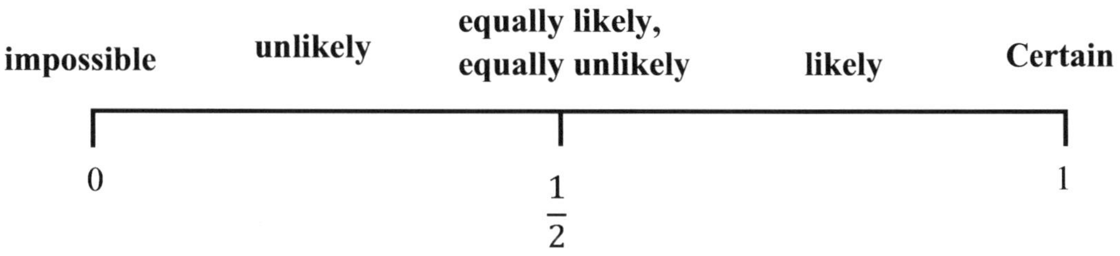

impossible **unlikely** **equally likely, equally unlikely** **likely** **Certain**

0 $\frac{1}{2}$ 1

Practice Test 4

Answers and Explanations

1) Answer: C

Divided 47 by 22: $\frac{47}{22} = 2.1363636 \ldots. = 2.1\overline{36}$

2) Answer: B

Day 3: $-0.64 \times \frac{3}{8} = -0.24$

Change price in days: $\big(86.95 + 0.58 + 0.76 + (-0.24) + (-0.64)\big) = 87.41$

Day 5: $88.29 - 87.41 = 0.88$

3) Answer: A

$1hour = 60\ min \rightarrow \frac{1}{4} \times 60 = 15\ min$

Rate: $\frac{\frac{3}{7}}{15} = \frac{x}{1} \rightarrow 15x = \frac{3}{7} \rightarrow x = \frac{3}{105} = \frac{1}{35}$

4) Answer: A

Probability $= \frac{\textit{number of desired outcomes}}{\text{number of total outcomes}} = \frac{2}{6+8+6+2} = \frac{2}{22} = \frac{1}{11}$

If an event has a 0 probability this means that it can never happen

If an event has a 1 probability it will certainly happen

If an event has a 0.5 probability, it has an equal chance of happening or not happening.

If an event has a probability between 0 and 0.5, then it is unlikely to happen.

If an event has a probability between 0.5 and 1, then it is likely to happen.

5) Answer: D

The pattern is: $5, -1, -7, -13, -19, -25, 31$

6) Answer: C

$54 \div 3 = 18$

$108 \div 6 = 18$

7) Answer: C

likely one-story house is: $0.42 \times 500 = 210$

likely of two-story house is: $0.52 \times 500 = 260$

8) Answer: D

Write the ratio and solve for x.

$\frac{1}{\frac{1}{6}} = \frac{x}{\frac{3}{5}}$ (Cross multiply) $\Rightarrow \frac{1}{6}x = \frac{3}{5} \Rightarrow x = \frac{18}{5}$

9) Answer: C

$3\frac{5}{9} \times \frac{-5}{9} = \frac{32}{9} \times \frac{-5}{9} = -\frac{160}{81} = -1\frac{79}{81}$

10) Answer: A

Hour: $x \rightarrow 26$ per hour: $26x$

Plus: add $(+)$, no more than: $\leq$; Then, $26x + 70 \leq 730$

11) Answer: B

By grid line: $d = 8$

Or distance for two points $(0,4), (0,-4)$: $d = \sqrt{(0-0)^2 + (4-(-4))^2} = 8$

$d = 8 \rightarrow$ Circumference $= \pi d = \pi(8) = 8\pi = 25.12$

12) Answer: D

2 hours equal 4 half hours

$4 \times 5° = 20° \rightarrow 7 - 20 = -13°$F

13) Answer: C

If the price of a printer is decreased by 38% then: $100\% - 26\% = 74\%$

$74\ \%\ of$ h $= 0.74 \times$ h $= 0.74$h

14) Answer: B

The median of a set of data is the value located in the middle of the data set. To find median, first list numbers in order from smallest to largest:

23, 23, 28, 30, 35, 38

Since there are an even number of items in the resulting list, the median is the average of the two middle numbers.

Median= $(28 + 30) \div 2 = 29$

15) Answer: A

Each hour is 60 minutes, so we have $4 \times 60 = 240$ minutes of workout time for Asher. We subtract off the jogging and playing volleyball time to get the time Asher playing billiards: $240 - 58 - 62 = 120$ minutes.

$$percent = \left(\frac{\text{part}}{\text{whole}}\right) \times 100 \rightarrow percent = \left(\frac{120}{240}\right) \times 100 \rightarrow percent = 50\%$$

16) Answer: B

$3(x + 5) + (2x + 11) + 59 = 180 \rightarrow 3x + 15 + 2x + 11 + 59 = 180$

$5x + 85 = 180 \rightarrow 5x = 95 \rightarrow x = \frac{95}{5} = 19 \Longrightarrow x = 19°$

$\angle B = 3(x + 5°) = 3(19° + 5°) = 72°$

17) Answer: C

The reflection of the point (x, y) across the y-axis is the point $(-x, y)$.

If you reflect a point across the y –axis, the y –coordinate is the same, but the x –coordinate is changed into its opposite.

Reflection of $(-6,4) \rightarrow (6,4)$

18) Answer: B

$-7x + 30 > -12$ (Subtract 30 from both sides)

$-7x > -42$ (Divide both side by -7, remember negative change the sign)

$x < 6$

19) Answer: D

1 L=1,000 mL

1 fl oz = 29.57 mL→ 1 fl oz =0.02957 L

$$118.28 \text{ L} = \frac{118.28}{0.02957} \times \frac{100{,}000}{100{,}000} = \frac{11{,}828{,}000}{2{,}957} = 4{,}000\ fl\ oz$$

20) Answer: A

Volume of a square-based pyramid: V $=\frac{1}{3}B.h$, where V is the volume and B is the area of the base. Then, V $=\frac{1}{3}\times 11.5\times 11.5\times 9=\frac{1{,}190.25}{3}=396.75$

21) Answer: A

Use formula of rectangle prism volume.

V = (length) (width) (height) ⇒ V $=15\times 6.8\times 12.7=1{,}295.4\ ft^3$

22) Answer: D

$28v\ +\ 1{,}120\ =\ 13{,}720$ (subtract 1,120 from both sides)

→ $28v=12{,}600$ (divide both sides by 28) → $v=\$450$ cost of each table.

23) Answer: B

Difference of temperature is: $|t_2-t_1|=|15-(-6)|=|15+6|=|-15-6|$

24) Answer: A

Supplementary angles are two angles with a sum of 180 degrees.

$\alpha+\beta=180°$ and $\beta=90°\Longrightarrow\alpha=180°-110°=70°$

complementary angles are two angles with a sum of 90 degrees.

$\alpha+\gamma=90$ and $\alpha=70°\Longrightarrow\gamma=90°-70°=20°$

25) Answer: D

A. Expenses in Year 2: 1,600 → $1{,}600+400=2{,}000\neq$ Year 3

B. Income in Year 1: 1,500 and $15\%\ of$ $1{,}500=225\rightarrow 1{,}500+225=$ $1{,}725\neq$ income of Year 5

C. Income's year 3, 4, and 5: $1{,}400+1{,}600+2{,}100=5{,}100$

Expense's year 2, 4, and 5: $1{,}600+1{,}300+1{,}500=4{,}400\neq 5{,}100$

D. Half of Year incomes 4: $\frac{1{,}600}{2}=800<1{,}100$ Expenses in Year 3

26) Answer: C

$(3n-9)-\frac{1}{3}(8-9n)+\frac{5}{3}=3n-9-\frac{8}{3}+3n+\frac{5}{3}=6n-10$

27) Answer: C

If you ever need to find the percentage of something you just times it by the fraction. So, all you need to do to figure this out is to find 35% of 13.06 which is approximately 4.57.

28) Answer: C

$\angle D = \angle E = b° \rightarrow FD = FE = 6cm$

Perimeter of triangle: $6 + 6 + 7 = 21\ cm$

Actual triangle: $21 \times 3\frac{1}{5} = 21 \times \frac{16}{5} = \frac{336}{5} = 67\frac{1}{5}\ cm$

29) Answer: B

The ratio of boy to girls is 4:5. Therefore, there are 5 girls out of 9 students. To find the answer, first divide the number of girls by 5, then multiply the result by 9.

$25 \div 5 = 5 \Rightarrow 5 \times 9 = 45$

30) Answer: D

A. does not consider that the manager's weight could be different from the mean weight and would change the value of the mean.

B. does not consider that the manager could be the same weight as the employee and that this weight could be the new mode.

C. does not consider that the nth heaviest employee (in the middle) could be different weight and adding a weight that is greater than either of these weights to the data set would change the value of the median.

D. corrects.

Practice Test 5

Answers and Explanations

1) Answer: C

Let y be the total amount paid.

We have been given that Peter bought 5 sandwiches that each cost the $12.95.

So, the cost of 5 sandwiches would be 5(12.95). He paid for 4 bags of fries that each cost the $2.45. So, the cost of 4 bags would be 4(2.45)

Then, the total cost of sandwiches and fries would be $y = 5(12.95) + 4(2.45)$

2) Answer: C

To get the answer to 10 over 27 as a decimal, we divide 10 by 27.

$\frac{10}{27} = 0.370370370 \ldots. = 0.\overline{370}$

3) Answer: B

Use the formula of circumference of circles.

Circumference = πd = 2π (r) = 14 π ⇒r=7

Radius of the circle is 7. Now, use the areas formula:

Area = πr^2 ⇒ Area = $\pi(7)^2$ ⇒ Area = 49 π

4) Answer: A

The volume of a triangular prism is the base times the height. $V = Bh$

Area of the base = $\frac{1}{2} b.h \rightarrow B = \frac{1}{2} \times 3 \times 4 = 6$

$V = B.h = 6 \times 6.5 = 39$ and we need two triangular prisms, then $2 \times 39 = 78$

5) Answer: A

$-4x - 1 < -7$, add 1 to both sides $-4x < -6$ divide each term by -4

If an inequality is multiplied or divided by a negative number, you must change the direction of the inequality, then $x > 1.5$

6) Answer: D

$(4+7)^2 + (4-7)^2 = (11)^2 + (-3)^2 = 121 + 9 = 130$

7) Answer: C

A. $(3 \times 1.85) + (3 \times 0.90) = 5.55 + 2.70 = 8.25 > 8$

B. $(5 \times 1.18) + (3 \times 0.76) = 5.90 + 2.28 = 8.18 < 8$

C. $(2 * 1.85) + (3 * 1.18) = 3.7 + 3.54 = 7.24 < 8$

D. $(5 \times 0.76) + (5 \times 0.90) = 3.8 + 4.5 = 8.3 > 8$

8) Answer: A

$0.16 \times x = 64 \rightarrow x = \frac{64}{0.16} = \frac{6400}{16} = 400$

$23\% of\ 400 = 0.23 \times 400 = 92$

9) Answer: C

$\sqrt{\frac{9x^7y^2}{49x}} = \sqrt{\frac{9}{49} \times \frac{x^7y^2}{x}} = \sqrt{\frac{9}{49}x^6y^2} = \frac{3}{7}x^3y$

10) Answer: D

Use Pythagorean Theorem: $a^2 + b^2 = c^2$

$6^2 + 6^2 = c^2 \Rightarrow 36 + 36 = c^2 \Rightarrow 72 = c^2 \Rightarrow c = \sqrt{72} = 6\sqrt{2} = 6 \times 1.4 = 8.4$

Perimeter of triangle $= a + b + c = 6 + 6 + 8.4 = 20.4$

11) Answer: D

Range= Largest Value – Smallest Value $\rightarrow$ R $= 8 - (-2) = 10$

12) Answer: C

To start, assign the variables to unknowns, known values to constants, and relate them by the relations between the variables and constants. Then

$9w + 80 = 215$

13) Answer: B

Use percent formula: part $= \frac{\text{percent}}{100} \times$ whole

Whole $= 8 + 5 + 7 = 20$

$8 = \frac{\text{percent}}{100} \times 20 \Rightarrow 8 = \frac{\text{percent} \times 20}{100} \Rightarrow 800 = \text{percent} \times 20 \Rightarrow \text{percent} = \frac{800}{20} = 40$,

Therefore United States win 40% of medals.

14) Answer: A

Multiply the price by the sales tax to find out how much money the sales tax will add, then Add the original price and the sales tax.

State A: $64 \times 0.0725 = 4.64$

$64 + 4.64 = 68.64$

State B: $48 \times 0.0825 = 3.96$

$48 + 3.96 = 51.96$

Then take the difference: $68.64 - 51.96 = 16.68$

15) Answer: A

18 physics teachers to 450 students is 18:450, 1:25

16) Answer: D

At least and Minimum – means greater than or equal to

At most, no more than, and Maximum – means less than or equal to

More than – means greater than

Less than – means less than

Then, $P \leq 30$

17) Answer: D

$\frac{16}{0.64}$ Multiply the numerator and denominator by 100: $\frac{1600}{64} = 25$

18) Answer: D

The ratio of boys to girls in Maria Club: 24:40 =3:5

The ratio of boys to girls in Hudson Club: 3:5

3:5 same as 15: 25.

So, there are 25 girls in Hudson club.

19) Answer: C

$\frac{3}{5} \times 4 = 2.4$

$\frac{3}{4} \times 60 = 45\ min$

$\frac{2.4}{45} = \frac{x}{60} \rightarrow 45x = 2.4 \times 60 \rightarrow x = \frac{144}{45} = 3.2$ Ounces

20) Answer: B

Supplementary angles are two angles that have a sum of $180°$

$line\ R: 180° - 132° = 48°, Line\ S: 180° - 117° = 63°$

then in the triangle: $180° - (63° + 48°) = 69°$

$line\ P: \theta° = 180° - 69° = 111°$

21) Answer: C

$3 \times (-7) = -21$

22) Answer: D

$\frac{282}{6} = 47$

$\frac{564-282}{12-6} = \frac{282}{6} = 47$

23) Answer: D

$36{,}850 \times 0.035 = 1{,}289.75$

24) Answer: D

The diagram shows a spinner with 8 sections. The probability land on letter is $\frac{5}{8}$

The spinner is spun 120 times, then the prediction is: $120 \times \frac{5}{8} = 75$ times.

25) Answer: B

A linear equation is a relationship between two variables, and application of linear equations can be found in distance problems.

$d = rt$ or distance equals rate (speed) times time.

$d = 1 \times 25 = 25$,then $(1,25), (2,50), (3,75)(4,100), \ldots$

26) Answer: C

average (mean) $= \frac{sum of terms}{number of terms} = \frac{(-25)+(-31)+(-15)+0+12+(-6)+16}{7} = \frac{-49}{7} = -7°F$

27) Answer: B

$m = 1 \rightarrow 4m - 1 = 4(1) - 1 = 3$

$m = 2 \rightarrow 4m - 1 = 4(2) - 1 = 7$

$m = 3 \rightarrow 4m - 1 = 4(3) - 1 = 11$

$m = 4 \rightarrow 4m - 1 = 4(4) - 1 = 15$

$m = 5 \rightarrow 4m - 1 = 4(5) - 1 = 19$

28) Answer: C

$-24 - 420d = -12(2 + 35d)$

29) Answer: A

Let's find the mode, mean (average), and median of the number of minutes for each school.

Number of Minutes for school 1: 50, 60, 60, 60, 80, 80, 90, 90, 110, 120

Mean(average) $= \frac{sum of terms}{number of terms} = \frac{50+60+60+60+80+80+90+90+110+120}{10} = \frac{800}{10} = 80$

Median is the number in the middle. Since there are an even number of items in the resulting list, the median is the average of the two middle numbers.

Median of the data is $(80 + 80) \div 2 = 80$

Mode is the number which appears most often in a set of numbers. Therefore, there is no mode in the set of numbers. Mode is: 60

Number of Minutes for school 2: 50, 60, 60, 90, 90, 100, 110, 120, 120, 120

Mean$= \frac{50+60+60+90+90+100+110+120+120+120}{10} = \frac{920}{10} = 92$

Median: $(90 + 100) \div 2 = 85$

Mode: 120

30) Answer: C

We often describe the probability of something happening with words like impossible, unlikely, as likely as unlikely, equally likely, likely, and certain. The probability of an event occurring is represented by a ratio. A ratio is a number that is between 0 and 1 and can include 0 and 1. An event is impossible if it has a probability of 0. An event is certain if it has the probability of 1

impossible **unlikely** **equally likely, equally unlikely** **likely** **Certain**

0 $\frac{1}{2}$ 1

Practice Test 6

Answers and Explanations

1) Answer: C

Divided 37 by 33: $\frac{37}{33} = 1.121212\ldots. = 1.\overline{12}$

2) Answer: B

Day 3: $-0.72 \times \frac{2}{3} = -0.48$

Change price in days: $\big(98.67 + 0.84 + 0.93 + (-0.48) + (-0.72)\big) = 99.24$

Day 5: $100.05 - 99.24 = 0.81$

3) Answer: A

$1hour = 60\ min \rightarrow \frac{1}{6} \times 60 = 10\ min$

Rate: $\frac{\frac{2}{5}}{10} = \frac{x}{1} \rightarrow 10x = \frac{2}{5} \rightarrow x = \frac{2}{50} = \frac{1}{25}$

4) Answer: C

Probability $= \frac{number\ of\ desired\ outcomes}{\text{number of total outcomes}} = \frac{1}{5+2+7+1} = \frac{1}{15}$

If an event has a 0 probability this means that it can never happen

If an event has a 1 probability it will certainly happen

If an event has a 0.5 probability, it has an equal chance of happening or not happening.

If an event has a probability between 0 and 0.5, then it is unlikely to happen.

If an event has a probability between 0.5 and 1, then it is likely to happen.

5) Answer: D

The pattern is: $7, -1, -9, -17, -25, -33$

6) Answer: C

$64 \div 4 = 16$

$128 \div 8 = 16$

7) Answer: C

likely one-story house is: $0.34 \times 300 = 102$

likely of two-story house is: $0.46 \times 300 = 138$

8) Answer: D

Write the ratio and solve for x.

$\frac{1}{\frac{1}{4}} = \frac{x}{\frac{2}{3}}$ (Cross multiply) $\Rightarrow \frac{1}{4}x = \frac{2}{3} \Rightarrow x = \frac{8}{3}$

9) Answer: C

$2\frac{7}{11} \times \frac{-7}{11} = \frac{29}{11} \times \frac{-7}{11} = -\frac{203}{121} = -1\frac{82}{121}$

10) Answer: A

Hour: $x \rightarrow 35$ per hour: $35x$

Plus: add (+), no more than: $\leq$; Then, $35x + 50 \leq 640$

11) Answer: B

By grid line: $d = 6$

Or distance for two points $(0,3), (0,-3)$: $d = \sqrt{(0-0)^2 + (3-(-3))^2} = 6$

$d = 6 \rightarrow$ Circumference $= \pi d = \pi(6) = 6\pi = 18.85$

12) Answer: D

2 hours equal 4 half hours

$4 \times 4° = 16° \rightarrow 9 - 16 = -7°F$

13) Answer: C

If the price of a printer is decreased by 38% then: $100\% - 38\% = 62\%$

$62\ \%\ of$ h $= 0.62 \times$ h $= 0.62$h

14) Answer: B

The median of a set of data is the value located in the middle of the data set. To find median, first list numbers in order from smallest to largest:

20, 20, 25, 27, 32, 35

Since there are an even number of items in the resulting list, the median is the average of the two middle numbers.

Median= $(25 + 27) \div 2 = 26$

15) Answer: A

Each hour is 60 minutes, so we have $3 \times 60 = 180$ minutes of workout time for Asher. We subtract off the jogging and playing volleyball time to get the time Asher playing billiards: $180 - 52 - 65 = 63$ minutes.

$$percent = \left(\frac{\text{part}}{\text{whole}}\right) \times 100 \rightarrow percent = \left(\frac{63}{180}\right) \times 100 \rightarrow percent = 35\%$$

16) Answer: B

$2(x + 13) + (3x - 6) + 45 = 180 \rightarrow 2x + 26 + 3x - 6 + 45 = 180$

$5x + 65 = 180 \rightarrow 5x = 115 \rightarrow x = \frac{115}{5} = 23 \Rightarrow x = 23°$

$\angle B = 2(x + 13°) = 2(23° + 13°) = 72°$

17) Answer: C

The reflection of the point (x, y) across the y-axis is the point $(-x, y)$.

If you reflect a point across the y –axis, the y –coordinate is the same, but the x –coordinate is changed into its opposite.

Reflection of $(-7,5) \rightarrow (7,5)$

18) Answer: B

$-8x + 40 > -16$ (Subtract 40 from both sides)

$-8x > -56$ (Divide both side by -7, remember negative change the sign)

$x < 7$

19) Answer: D

1 L=1,000 mL

1 fl oz = 29.57 mL→ 1 fl oz =0.02957 L

$$59.15 \text{ L} = \frac{59.15}{0.02957} \times \frac{100{,}000}{100{,}000} = \frac{5{,}915{,}000}{2{,}957} \approx 2{,}000\ fl\ oz$$

20) Answer: A

Volume of a square-based pyramid: V $=\frac{1}{3}B.h$, where V is the volume and B is the area of the base. Then, V $=\frac{1}{3}\times 10.5\times 10.5\times 7=\frac{771.75}{3}=257.25$

21) Answer: A

Use formula of rectangle prism volume.

V = (length) (width) (height) $\Rightarrow$ V $= 17\times 8.4\times 11.25=1{,}606.5\ ft^3$

22) Answer: D

$25v + 1{,}350 = 12{,}850$ (subtract 1,350 from both sides)

$\rightarrow 25v = 11{,}500$ (divide both sides by 25) $\rightarrow v = \$460$ cost of each table.

23) Answer: B

Difference of temperature is: $|t_2 - t_1| = |17-(-9)| = |17+9| = |-17-9|$

24) Answer: A

Supplementary angles are two angles with a sum of 180 degrees.

$\alpha+\beta = 180°$ and $\beta = 90° \Longrightarrow \alpha = 180° - 90° = 90°$

complementary angles are two angles with a sum of 90 degrees.

$\alpha+\gamma = 90$ and $\alpha = 90° \Longrightarrow \gamma = 90° - 90° = 0°$

25) Answer: D

A. Expenses in Year 2: 1,700 $\rightarrow 1{,}700 + 200 = 1{,}900 \neq$ Year 3

B. Income in Year 1: 1700 and $20\%\ of\ 1{,}700 = 340 \rightarrow 1{,}700 + 340 =$ $2{,}040\ \neq$income of Year 5

C. Income's year 2, 3, and 4: $1{,}900 + 1{,}600 + 1{,}800 = 5{,}300$

Expense's year 2, 4, and 5: $1{,}700 + 1{,}600 + 1{,}800 = 5{,}100 \neq 5400$

D. Half of Year incomes 4: $\frac{1{,}800}{2} = 900 < 1{,}400$ Expenses in Year 3

26) Answer: C

$(2n-8)-\frac{1}{2}(9-4n)+\frac{5}{2} = 2n-8-\frac{9}{2}+2n+\frac{5}{2} = 4n-10$

27) Answer: C

If you ever need to find the percentage of something you just times it by the fraction. So, all you need to do to figure this out is to find 45% of 11.08 which is approximately 4.99.

28) Answer: C

$\angle D = \angle E = b° \rightarrow FD = FE = 7cm$

Perimeter of triangle: $7 + 7 + 8 = 22\ cm$

Actual triangle: $22 \times 2\frac{1}{4} = 22 \times \frac{9}{4} = \frac{198}{4} = 49\frac{1}{2}\ cm$

29) Answer: B

The ratio of boy to girls is 2:3. Therefore, there are 3 girls out of 5 students. To find the answer, first divide the number of girls by 3, then multiply the result by 5.

$18 \div 3 = 6 \Rightarrow 6 \times 5 = 30$

30) Answer: D

A. does not consider that the manager's weight could be different from the mean weight and would change the value of the mean.

B. does not consider that the nth heaviest employee (in the middle) could be different weight and adding a weight that is greater than either of these weights to the data set would change the value of the median.

C. does not consider that the manager could be the same weight as the employee and that this weight could be the new mode.

D. corrects.

"End"

Made in the USA
Las Vegas, NV
12 April 2024